Underground Disposal of Coal Mine Wastes

A Report to the NATIONAL SCIENCE FOUNDATION
by the
Study Committee to Assess the Feasibility of Returning Underground
Coal Mine Wastes to the Mined-Out Areas

Environmental Studies Board/Board on Energy Studies

National Academy of Sciences/National Academy of Engineering

NATIONAL ACADEMY OF SCIENCES
Washington, D.C., 1975

Library of Congress Catalog Card No. 74-32597
International Standard Book No. 0-309-02324-6

Available from
Printing and Publishing Office, National Academy of Sciences,
2101 Constitution Avenue, N.W., Washington, D.C. 20418

Printed in the United States of America

STUDY COMMITTEE TO ASSESS THE FEASIBILITY OF RETURNING UNDERGROUND COAL MINE
    WASTES TO THE MINED-OUT AREAS

CHARLES FAIRHURST, University of Minnesota, Minneapolis, Chairman
JOSEPH P. BRENNAN, National Coal Association, Washington, D.C.
THOMAS V. FALKIE, Pennsylvania State University, University Park, Pennsylvania;
    U.S. Bureau of Mines, Washington, D.C.
LAWRENCE A. GARFIELD, Kellogg Corporation, Englewood, Colorado
JAMES E. GILLEY, U.S. Bureau of Mines, Washington, D.C.
RICHARD E. GRAY, General Analytics, Monroeville, Pennsylvania
WALTER N. HEINE, Pennsylvania Department of Environmental Resources, Harrisburg
DAVID R. MANEVAL, Appalachian Regional Commission, Washington, D.C.
J. DAVITT McATEER, United Mine Workers of America, Washington, D.C.
WILLIAM N. POUNDSTONE, Consolidation Coal, Pittsburgh, Pennsylvania

Special Advisors

STEVEN L. CROUCH, University of Minnesota, Minneapolis
LEE W. SAPERSTEIN, Pennsylvania State University, University Park, Pennsylvania

For the NATIONAL ACADEMY OF SCIENCES/NATIONAL ACADEMY OF ENGINEERING:

GENEVIEVE ATWOOD, Staff Officer
FRANCES J. MOODY, Project Secretary

Government Liaison

JOHN J. MULHERN, Environmental Protection Agency, Washington, D.C.
STEVEN GAGE, Council on Environmental Quality, Washington, D.C.

National Science Foundation Liaison

RAYMOND JOHNSON, National Science Foundation, Washington, D.C.

COMMISSION ON NATURAL RESOURCES

GORDON J.F. MacDONALD, Dartmouth College, Hanover, New Hampshire, Chairman
WILLIAM C. ACKERMANN, Illinois State Water Survey, Urbana
JOHN E. CANTLON, Michigan State University, East Lansing
RONALD FREEDMAN, University of Michigan, Ann Arbor
HAROLD L. JAMES, U.S. Geological Survey, Menlo Park, California
JOHN J. McKETTA, University of Texas, Austin
EMIL MRAK, University of Texas, Austin

TABLE OF CONTENTS

Table of Contents (Continued)

Table of Contents (Continued)

TABLES AND FIGURES

Tables and Figures (Continued)

Tables and Figures (Continued)

Underground Disposal of Coal Mine Wastes

CHAPTER 1
FINDINGS AND CONCLUSIONS

The production of waste is an inevitable consequence of any industrial process or, indeed, any process involving a change of physical or chemical state. Until recently, little concern had been expressed for disposition of waste beyond the waste basket. "Get it out of sight" has been the undefined philosophy of waste disposal throughout the history of industrial development.

Increasing demands for more material comforts, recreation and travel by growing populations have led to unprecedented production of waste and also to a perception of the problems associated with waste disposal. Waste is no longer out of sight. The public, aware of the environmental degradation that has taken place in the past and seeing that it continues to occur demands a halt to it; but the same public is reluctant to forego the comforts that lead to the waste. The goals of modern technology have in effect been redefined: material comforts are to be produced, but without lasting adverse effect on the environment. The engineer must determine how to provide material comforts at minimum cost, both economically and environmentally.

Disposal of coal mine wastes in the United States is an excellent example of this situation. The public demands that coal be produced to meet the nation's energy needs, but is understandably appalled by the sight of burning waste piles and haunted by the hazards portended by Aberfan and Buffalo Creek (see section 2.3). Underground disposal of coal mine waste has been proposed as a means of attaining these demands and yet avoiding the problems associated with waste piles. Relatively little attention seems to have been paid to the effect of underground disposal on the underground mining environment.

Coal mine waste has been stowed underground in Europe for economic reasons, but not with waste disposal as the primary motivation. The economic justifications for backfilling (principally maximum recovery of the coal resource and subsidence control) have changed, and the use of backfilling has now declined sharply--although the environmental concerns for land use and waste management have multiplied. European experience indicates that underground waste disposal is technically feasible, although significant differences in European mine conditions, land values, labor rates, and environmental pressures must be recognized when discussing the feasibility of waste disposal in the United States.

Returning wastes from underground coal mines to the underground, the only disposal option that this committee examined at length, is one possibility for environmentally acceptable disposal. In making a decision on optimal disposal

1

solutions, underground disposal should be examined in the context of the full resource problem, and compared to the alternative disposal options and the respective costs and benefits of each. The optimal disposal solution will likely differ from mine to mine in response to a wide variation in inherent physical, economic, social, and environmental conditions of individual mines. Three inherently different waste disposal problems should be addressed separately and the cost for each assumed in different ways (Table 1.1).

Higher production costs due to changes in coal mining methods to accommodate underground disposal will have a differential economic effect not only on the price of coal as related to other energy sources, but also on different mining methods (i.e., underground or surface), and on individual underground operations depending on the facility with which they adapt to the criteria. The quantifiable economic costs of underground disposal could be reflected in higher production costs, losses in productivity, losses in coal availability, or an accelerated shift to surface mining methods. Those considerations must be carefully weighed against the benefits to the surface environment and greater public safety that would be attendant on returning the waste underground.

However, underground waste disposal must not become a variant of the "get-it-out-of-sight" mentality. Environmental concerns for the health and safety of men in the underground working environment, and the possible long-term environmental consequences of backfilling--such as the degradation of groundwater quality--must be considered against the long-term advantage of disposing of waste underground.

1.1 CONCLUSIONS

1. Underground waste disposal is technologically feasible in most instances and is one of a number of ways of dealing with waste in an environmentally satisfactory manner. Underground waste disposal is technologically feasible in most abandoned mines and in active mines where longwall mining or partial recovery (a method of room and pillar mining) is practiced. Approximately 2 percent of U.S. underground coal mines have longwall panels and approximately 70 percent practice partial recovery. The economic feasibility of backfilling

TABLE 1.1

Type of Coal Waste Disposal Problem	Financial Responsibility
1. Existing and abandoned waste piles inherited from past abuses	1. Society in general, i.e., the federal government
2. Waste production from active mines using "unacceptable" disposal schemes	2. Some form of supplemental money-- be it partial government subsidy or increase in the price of coal
3. Acceptable disposal schemes designed for new mines	3. The producer--the price of the product for the competitive supply-demand marketplace

varies with specific sites and conditions. Underground waste disposal is not economically feasible in many operations, especially in currently operating mines. In future mines designed for underground disposal, or under different economic constraints, the economic feasibility of underground waste disposal could improve.

2. Serious unresolved concerns relating backfilling to the health and safety of mine personnel deserve further consideration. Underground stowing in active mines increases the number of men exposed to an already dangerous working environment. In addition, there may be health and safety hazards related directly to the backfilling process. The committee could not establish that an increase in health or safety hazards would be contingent upon the backfilling process, but it did recognize that special provisions in current mine health and safety laws and regulations would have to be made in order to permit its use. Underground disposal of waste eliminates a potential hazard to workers and area residents which can exist when improper techniques of surface disposal are used.

3. Technology for safe and nonpolluting surface waste disposal exists in most areas and offers a viable alternative to underground disposal provided stringent regulations for construction and maintenance of surface disposal areas are strictly enforced.

4. More consideration should be given to coal mine waste as a potential resource.

5. Costs of both underground and surface disposal will vary with the specific conditions of each mine. If underground disposal is deemed socially desirable, a means should be found to prevent a significant economic imbalance among currently operating mines that could result from such a change in mining operations. Safe and nonpolluting disposal is the responsibility of the operator. If regulations are imposed that exceed this obligation, society must recognize its responsibility to avoid inequitable distribution of added costs to some operators.

The charge given the committee was to consider the current technological and economic feasibility of backfilling with mine wastes. The effect of different disposal systems on land use, their long-term cost to the environment, and their psychological and sociological impacts on the community were outside the committee's delineated task. When decisions affecting refuse disposal are made, these factors must be recognized as being of equal importance as the technical-economic concerns.

1.2 GENERAL FINDINGS OF THE COMMITTEE

1. A coal mine operator is responsible for disposing of the wastes from his operation in a manner that is not hazardous to the health or safety of the public or mine workers, not harmful to the environment, not aesthetically unacceptable, and not detrimental to other regional land uses. In the past the coal industry did not, and even today some operators do not accept this responsibility. Nevertheless, forward looking companies have taken actions to meet it.

2. Although this study is directed primarily toward the feasibility of backfilling in active mines, the serious problems of orphaned, unsafe waste piles should not be neglected.

3. Underground disposal of coal mine wastes other than fines* is not practiced in the United States. Without direct experience in active U.S. mines, the committee cannot recommend underground disposal over surface disposal methods or vice versa, but rather urges that underground disposal be considered as one possible means of disposing of wastes in an acceptable manner. It should be considered in the context of all other acceptable options and the relative economic, technological, environmental, social, and health or safety considerations of each. Such considerations would have taken the committee far beyond its charge: to assess the technological and economic feasibility of returning coal mine wastes underground.

4. Underground disposal of coal mine waste is one of several disposal methods in active mines, and can also be considered a means of subsidence control in abandoned mine workings or sections of working mines where pillar support appears inadequate. Backfilling of underground mine openings significantly reduces surface damage from subsidence by lending lateral support to pillars and by limiting the volume of the voids. Backfilling does not completely eliminate subsidence but reduces it about one-third compared to caving non-backfilled mines. Since the refuse produced in the coal cleaning process in the United States is currently only 25 percent of the volume of the raw coal, total filling of mine openings with waste from the mining operation alone is not possible. European experience indicates that in multiple seam mining there are more economical and effective ways of compensating for or reducing damage from subsidence than backfilling.

5. Of the proposed schemes for utilization of waste in order to dispose of it, only the use of waste for highways and construction fills could use substantial quantities. Landfill and other construction projects in areas of refuse piles should be designed and located to use as much coal mine waste material as possible. However, the quantities of current production, plus existing and orphaned waste material, are so great that it is unlikely that construction will prove to be the answer to the problem of either orphaned bank disposal or active mine waste disposal.

1.3 CONCLUSIONS REGARDING TECHNICAL FEASIBILITY

1. Underground backfilling of coal mine wastes is technically feasible in most active and abandoned mines using current technology. This assertion is based upon experience in active coal mines in Europe and upon limited back-filling of abandoned coal mines in the United States. Although conditions in these mines may differ from those of active underground coal mines in the United States, some of the technology developed is transferable. In light of the differences in mining conditions between Europe and the United States, some practical U.S. experience with backfilling should be developed if a meaningful evaluation of the economic and social costs of various practices is to be made.

---

*See Appendix A, Glossary of Terms for definitions of mining terms.

2. The feasibility of backfilling underground coal mines must be assessed for each specific mine site. The most economical and technically effective underground disposal schemes will be those developed for new mines in which the method of disposal is designed as part of the mining method. Conversely, it is difficult to achieve the optimum in economic and technical disposal in existing active mines; and it has been demonstrated that the highest costs are those by backfilling abandoned mines. Because each mine site is distinctive, there is no categorical answer to the desirability of underground backfill over any other disposal system. The geological conditions of the coal deposit, the mining transport mechanism, the proposed disposal scheme, and also the local societal values placed upon the social and environmental consequences of mining and waste disposal, must all be considered in assessing the merits of various alternative disposal systems.

3. Advantages of stowing coal waste underground include the elimination of the environmental, health, safety and social problems associated with surface disposal. Where backfilling would provide lateral support to pillars or significantly reduce the mine voids, underground disposal could also reduce subsidence (see Chapter 3 and Appendix B). The disadvantages of underground disposal relate to problems associated with the health and safety of miners underground, the loss of potentially recoverable coal, losses in productivity, and possible environmental problems associated with the construction of roads and injection boreholes on the surface, and groundwater contamination from leaching of the backfilled material. All of these problems may not exist at a given operation and all may be significantly reduced given sufficient money and time. There may be potential industrial benefits to be derived from disposing of coal wastes underground if wastes from other sources--such as coal-fired, power plant fly ash or sulfate sludges--can be disposed of at the same time. At the present time, the following three systems appear to be the most applicable for backfill in active mines in the United States: (a) pneumatic stowing machines available from foreign sources to backfill behind longwall miners. Waste material would be introduced into the mine from boreholes on the surface; (b) backfill of partially mined areas from the surface, introducing the slurry via borehole pipelines into the mine openings; (c) pneumatic backfilling of partially mined areas of room and pillar operations.

4. A method of mining with concurrent backfilling for modern room and pillar operations has not been developed for active coal mines at this time. Such a mining method, if developed, would have widespread applicability in the United States and be preferable to the room and pillar backfill systems listed below.

5. At the present time, the following two systems appear to be the most feasible for backfilling abandoned mines in the United States--depending upon the stability of the existing pillars and the degree of flooding in the mines: (a) controlled flushing (where safe); (b) blind flushing by hydraulic methods.

6. Consideration should be given to a combination of disposal methods; in particular, the disposal of the larger size fractions of waste on the surface using environmentally acceptable methods which emphasize prevention of spontaneous fires, combined with underground disposal of the fine fraction.

1.4 ECONOMIC FEASIBILITY

The costs of handling mine waste in an environmentally acceptable manner must be considered part of the normal expense of production for active coal mines.

1. Economically, it does not appear that the cost of underground stowage is necessarily prohibitive in all mines, although underground disposal is considerably more expensive than current methods of surface disposal. The economic feasibility of waste disposal underground must be evaluated for each mine operation in the light of the competitive position it holds in the industry and the competitive position of coal in the energy market.

2. The cost of underground disposal will vary widely from mine to mine and in some cases may be prohibitive. Advances in waste disposal technology may significantly reduce the costs of returning coal mine wastes underground in future mines, but underground disposal may be expected to be more expensive than current methods of surface disposal. If wastes are to be placed underground, economic and legal incentives will no doubt have to be provided or current surface disposal methods will have to be further regulated or made less economically attractive.

3. Data are available for the costs of backfilling for subsidence control in the United States and in Europe, but figures for backfilling strictly for waste disposal can only be estimated. Some, but not all of the costs for subsidence control are directly transferable to costs for disposal. Experience in Europe has shown that backfilling is expensive and that there are less expensive ways of limiting subsidence damage than by means of underground disposal of waste. However, the unfavorable economics of backfilling for *subsidence control* in Europe does not mean that backfilling for *disposal* will be economically prohibitive in the United States. The committee estimates that at minimum, under unusually favorable mining conditions, and average conditions of waste production that it would cost at least $1-$4 per ton of coal to stow wastes underground.

4. If the return of coal mine waste underground should be required of all active operations, the added costs to underground mines in competition with surface operations could force closure of some mines. However, increased energy prices suggest that industry-wide improvements in technology can be afforded at this time although some operators would be hurt more than others.

5. The committee tried to determine the cost of physically placing refuse in active and abandoned mines. This cost was compared with present costs of surface disposal. This method may be inadequate because such a cost analysis compares anticipated costs with present operating costs, assumes that current disposal practices are "environmentally acceptable," and, most important, does not consider the social costs incurred by current and past surface disposal practices or by underground disposal.

1.5 SOCIAL AND LEGAL FACTORS

This report considers only one alternative to current disposal tech-
niques--underground disposal. Increased public concern for the degrading
effects of past surface disposal practices may provide the impetus for legislation to return coal waste to mined out areas.

1. [At the present time, it is not legally required that coal mine wastes be
disposed underground. Responsibility for the enforcement of laws, for regu-
lating methods of disposal, and for monitoring of environmental effects of
disposal, is dispersed among several federal and state agencies.] A more co-
herent policy concerning waste disposal should be formulated which delineates
enforcement responsibilities as well as the criteria for disposal.

2. [It is clear that a distinction exists between the problems of backfilling
in active and abandoned mines. It probably will require action by govern-
mental agencies to eliminate or abate the environmental problems resulting
from abandoned coal waste piles.] There is no technical reason why the waste
from orphan piles cannot be placed into underground mines. However [deter-
mination of the legal ownership of the piles, and the willingness of society
to accept the burden of cleanup may present obstacles to this possibility.]

3. There is a need for further legal regulation of waste disposal. Two
possible approaches may be considered: (a) to prescribe the methods for
proper disposal, or (b) to prescribe the goals of proper disposal.

  a. In prescribing methods, thorough analysis of long-term damage of
surface disposal may reveal the benefits of backfilling to be large enough to
offset the costs. In this approach backfilling of all mine waste would be
required at active mine operations after sufficient time has elapsed to develop
the technology and manufacture the hardware required, to resolve concerns for
the health and safety of mine personnel, to develop applicable health and
safety standards for individual mines, and to reformulate current regulations
to include special provisions for backfilling.

  b. The goals approach is endorsed by the majority of the committee.
Regulations would be developed to ensure safe, nonpolluting disposal of coal
mine waste by either surface or underground methods. This approach would
prohibit long-term degradation of the land, the water or the air, but it would
not prescribe a specific disposal method for a mine operation. The operator
could treat the waste at his option, either by underground stowage or proper
surface disposal. Rules would regulate discharge, stability, safety, and the
aesthetics of disposal.

1.6 RESEARCH NEEDS

In completing the assignment to examine the feasibility of underground
disposal of coal mine wastes in the United States, the committee determined

that adequate demonstration of this method of disposal does not exist on this continent nor is detailed information on foreign practice readily available. In view of these deficiencies, the committee recommends:

1. The establishment of a cooperative international program with European engineers to collect and disseminate information on the latest advances in technology including the machines and mining techniques used in surface and underground disposal of coal mine waste.

2. The operation of at least two underground waste disposal demonstration projects at active mines in the United States which may be considered representative of (a) complete extraction mining and, (b) partial recovery mining. The projects should be directed to determine the hazards to health and safety that are attendant on underground waste disposal and how the hazards may be reduced or eliminated. Emphasis should be on developing ways to enhance mine safety and to minimize damage to the overlying surface. The projects should be full-scale, since pilot demonstrations can have irreconcilable limitations in establishing costs and technical feasibilities. Maximum participation of all interested groups to effect cost and time savings might be achieved through joint project sponsorship by the operating mining company; the local union; federal, state and local governments; and community representatives.

CHAPTER 2
STATEMENT OF THE PROBLEM

SUMMARY

Coal mine waste, already a serious problem due to that accumulated from past coal mining operations, presents a serious environmental and social stress, especially on Appalachia. With increased coal production, waste disposal will become an even greater problem.

Approximately 25 percent of the coal extracted from underground mines in the United States is rejected as waste and disposed on the surface near coal preparation plants. The amount that has accumulated over the past 200 years is considerable. In the eastern coal fields alone there are at least 3,000-5,000 sizable active and abandoned waste piles and impoundments which cumulatively contain over 3 billion tons of refuse.

The rate at which these mine wastes accumulate varies according to the nation's total production of coal, the proportion of raw coal cleaned, the prevailing market preference for different qualities of coal and the specific mining and cleaning methods used. With current pressures for increased production of coal and the desire for coal of higher quality to enable air quality standards to be met, it is likely that more coal waste will be produced annually in the next few decades than has been produced in the past.

These waste piles produce acute social as well as environmental problems. They have been the direct cause of the death of mine workers and local inhabitants, and the source of severe local air pollution and stream and groundwater pollution, as well as having had regional, economic, psychological and sociological impacts. Proper surface or underground disposal can minimize such problems at future sites but inactive piles continue to present a long-term negative impact to the region and its people.

Alternatives to disposal have been suggested to eliminate the problems of past and future waste. These include the recycling of old waste piles for their fuel, and the use of waste for construction fill, for concrete aggregate, road base, anti-skid materials, or for brick materials. Unfortunately, the quantity of waste that can be utilized by such means *is in question*. Waste is a potential asset, but it rarely is located where it can be most useful. Thus, the disposal of coal waste is already a serious problem that will tend to grow in magnitude--not to disappear.

2.1 MAGNITUDE OF THE PROBLEM

"At present many of the collieries in the anthracite region of
Pennsylvania are utilizing refuse from the old culm banks, which formerly
were prominent features in the landscape, and are flushing the fine waste
underground so that pillars may be reduced in size or removed. As a
result the banks are now disappearing and soon will be gone." (U.S.
Department of the Interior, 1913b)

Refuse Production Past and Present

Most coal seams contain thin partings, lenses or inclusions of slate,
shale, siltstone, or claystone which are recovered as part of the coal mining
operation. Before current mine mechanization this waste rock was removed from
the coal by hand underground or on the surface by hand or by screening. After
World War II the increased use of mechanical mining equipment made it impractical to selectively reject waste material underground. Today, marketable
clean coal is separated from the noncoal waste at preparation plants by mechanical and physical methods. This waste material is disposed on the surface by
conveyor belt, truck, gondolar car, aerial tramway, or other means onto waste
piles that range in size from a few acres, to piles hundreds of feet high and
over a mile long. The very fine waste material generated in coal processing
generally is collected in settling basins which upon filling are usually
abandoned or, upon drying, are disposed on the coal waste pile.
  The rate of accumulation of coal mine wastes is related to mining methods.
The 30-year period since 1940 shows a steady increase in waste by nearly 10
fold (Figure 2.1a and b). A significant portion of this increased waste
accumulation is attributable to modern mining equipment which removes approximately three times more roof and floor rock than was removed by hand methods.
Recent statistics indicate that 25 percent of mined coal is rejected and has
to be disposed as waste (U.S. Department of the Interior, 1973).
  Although some of the refuse accumulated over the last 200 years has been
recycled for its fuel content, used as landfill or utilized for some other
purpose most of it remains in waste piles. The actual number and volume of
these piles has not been determined although in 1966, 1968 and 1972 the Bureau
of Mines inventoried waste piles and impoundments in different sections of the
country. In 1968, 961 large refuse banks across the country were studied to
determine those which posed public health and safety problems of sliding,
burning or impounding water. Over half posed some form of problem (U.S.
Department of the Interior, 1968). In a survey conducted shortly after the
Buffalo Creek disaster in 1972, an attempt was made to inventory all waste
impoundments and piles in order to categorize the safety and health hazards
associated with them. In the Anthracite region alone, 812 waste piles were
inventoried of which 631 were inactive, 27 had been burned, 28 were burning,
109 impounded water and 10 posed a potential safety hazard (U.S. Department of
the Interior, unpublished, c). There are approximately 70,000 abandoned or
inactive underground coal mines in the United States and approximately 2,000

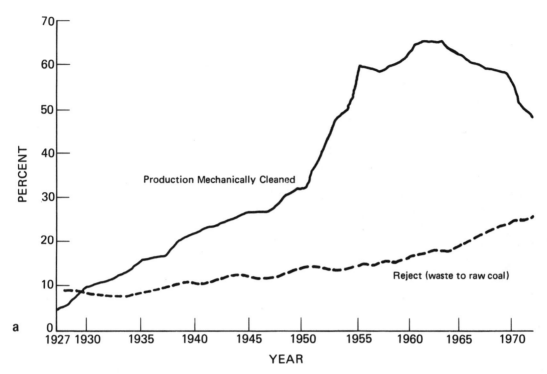

FIGURE 2.1a Trends in portion of total production of coal mechanically cleaned and average percentage reject in U.S.

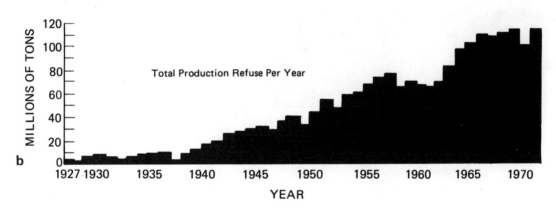

FIGURE 2.1b Total production refuse per year.

active underground coal mines. Most of these mines at some time have had a waste pile associated with them (see e.g., Figures 2.2a, b, and c; and Figure 2.3). The total number and volume of sizable active and abandoned waste piles and impoundments in the eastern coal fields alone is conservatively estimated at 3,000-5,000 cumulatively containing over 3 billion tons of refuse (U.S. Department of the Interior, unpublished a, b, c, 1964, 1969b, 1971b, 1972a, 1973b).

FIGURE 2.2a, b, and c   Progressive stages of extinguishment of Marvin Bank, Scranton, Pennsylvania

FIGURE 2.2a   The bank in 1968.

FIGURE 2.2b   The bank in September, 1969 after an initial $1,750,000 project of extinguishment.

FIGURE 2.2c   The bank in September, 1970 after the second stage of extinguishment costing $1,525,000. Note the proximity of the bank to Interstate Highway I-81. The bank contains 2,015,000 tons of waste. The cost of extinguishment was $0.75-$1.00/ cubic yard of waste.

FIGURE 2.3 Environmentally acceptable surface disposal, Bethlehem Mines Corporation, Ellsworth, Pennsylvania, Mine 60. Note the results of seeding and planting on the benches of the pile in the foreground. New mine waste pile on the horizon (Bureau of Mines).

Projection and Trends of Future Waste Production

Coal resources represent nearly 80 percent of the fossil fuel reserves in the United States. The portion of total energy demand that has been met by coal has diminished over the past 60 years. The actual amount of coal consumed has remained constant since 1945 although it will probably double by the year 2000 (Figures 2.4 and 2.5). Should coal meet a portion of the energy demand projected to be met by petroleum and natural gas, the demand for coal may double by 1985 and quadruple by the year 2000.

Due to changes in mining methods and cleaning processes over the past thirty years, the rate of accumulated waste has increased ten times more rapidly than the rate of coal production. However, market preference over the past ten years for uncleaned coal is a countervailing factor which has tended to decrease the rate of waste accumulation. Due to economic factors such as lower transportation rates, electric utilities have bought uncleaned coal to burn in pulverized fuel stokers. The waste is accumulated as fly ash after the mixture has been burned. Because power plants consume over half the coal produced in the United States, a shift in their consumptive patterns strongly affects the production of mine waste at preparation plants. In the past ten years, the power plants' preference for uncleaned coal has led to a decrease in the percentage of total coal production which is mechanically

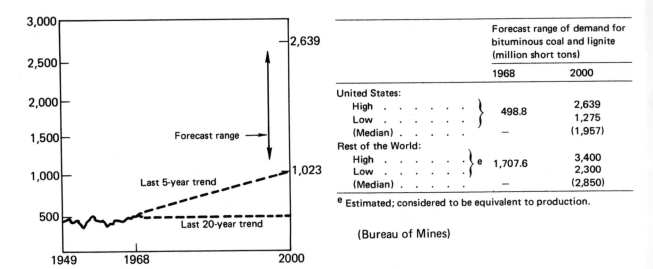

FIGURE 2.4 Comparison of trend projections and fore-
casts for bituminous coal and lignite demand.

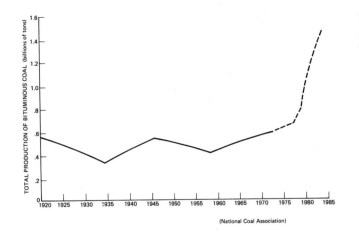

FIGURE 2.5 Past production
and projected production of
bituminous coal and lignite
through 1985.

cleaned from almost 65 to 50 percent. Should the trend toward the use of uncleaned coal continue, current rates of coal refuse production at the mine
site and preparation plant will not increase as rapidly as overall coal production and the quantity of fly ash will increase proportionately faster.
However, recent trends in consumption patterns of power plants in Pennsylvania
indicate that power plants, in their efforts to meet clean air requirements,
increasingly want coal with lower sulfur content and have wanted cleaned coal
(Hambright, personal communication, 1973). Thus, it is difficult to predict
whether the ratio of waste production will change significantly over the next
decade. Certainly the quantity of waste will increase with increased production of coal. The committee estimates that the net effect of the impact on

waste production by increased production of coal, changing mining methods, changing cleaning methods, increased emphasis on clean fuels that tend to increase waste products and the impact of the use of uncleaned underground and surface-mined coal that tends to decrease the accumulation of coal mine waste will result in a net increase in waste production from approximately 100 million tons at the present time to 200 million tons by 1980.

2.2 PROPERTIES OF WASTE MATERIAL

Two types of waste are generated by the coal preparation plants: coarse waste (particle cross-section greater than one millimeter) and fine waste (less than 1 millimeter). Fines constitute approximately 10 percent of coal mine waste. The properties of coal waste vary according to the mineralogical constituents of the waste rock contained in noncoal bands within the coal, the composition of the adjacent strata, the method of mining the coal, the method and efficiency of the cleaning operation, and the quality of the coal and the market for which it is cleaned. The properties of mine waste from fresh waste piles is compared to those of mine wastes from a weathered waste pile and those of bituminous coal in Table 2.1.

Laboratory examinations of the physical properties of coal mine waste have resulted in several generalizations covering waste piles (U.S. Bureau of Mines, 1974). The grain-size distribution within a pile tends to be uniform although the size of the particles varies considerably. The shear strength and permeability of waste piles also are relatively uniform. Specific gravity and bulk density vary among banks but values are generally constant for a particular waste pile. The variability among piles appears to be due to the method of cleaning rather than geologic conditions. It appears that uniform high density and low permeability are a function of the manner in which the waste material is placed. The stability of a pile is a function of the soil or rock which forms the foundation of the pile, the properties of the waste material and the manner in which the waste material is placed. Although an operator cannot control the strength of the natural strata on which he builds the pile, he can design the waste pile accordingly. In addition, as exemplified by British experience,

"The brittleness of fills, on the other hand, lies within the controls of the engineer. Waste materials, which in their loose state have led to the most disastrous or potentially dangerous flowslides, are in general excellent construction materials if compacted at the correct water content. Only slurry lagoons present a less tractable problem, in particular, where they are normally consolidated and are permitted to remain fully saturated (as is almost inevitable in a temperate climate)." (Bishop, 1973, p. 374)

TABLE 2.1 Comparison of the Physical Characteristics of Coal, Fresh Coal Waste and Weathered Waste

	Clean Coal (95% of cleaned coal is bituminous)	Fresh Waste (Active Piles) (Content very dependent on characteristics of roof material: stability dependent on disposal practice)	Weathered Waste (Old Piles) (dependent on coal content and how much coal has burned out)
Specific gravity	1.23-1.72	varies considerably from pile to pile range 1.6-2.7 average 2.2	varies considerably, primarily by the coal content range 1.4-2.7 average 2.0
Size	varies a lot dust--6 in.	The size range is variable, but the pile is usually well-graded Coarse: normally 4 in., rarely 8 in.	More fines and more fine coal than fresh refuse
Moisture Content	range 1-40% average 3-6%	Dry refuse 5-10% Cleaning plant refuse 10-40% Slurries and sludge 25-70% (30% solids)	10-20%
Carbon & Volatile Content (Dry Basis)	80-95%	range 7-25%	10-45%
BTU per pound	10,000-15,000	1,500-6,000	Higher BTU in general than fresh refuse--3,000-10,000
Sulfur Content	1-5%	range 3-15% average 5-10%	Less sulfur the longer the refuse is leached
Permeability	in situ: $10^{-1}-10^{-3}$ cm/s	Uncompacted--high permeability Compacted--$10^{-2}-10^{-4}$ cm/s	High Permeability Varies with the configuration of the pile

TABLE 2.1 (Continued)

	block of coal	Angle of drained shearing	average drained shearing
Shear Strength	block of coal 200-1000 psi average 700	Angle of drained shearing resistance 25.5°-41.5° average 30° compacted: higher angle	average drained shearing strength 30° saturated average 11°
Compressive Strength	500-6000 lb/in.²	100-500 psi compacted	50-150 psi
Drained Cohesive Strength	Not applicable	zero	zero
Ash	range 3-12% average 8%	- - - -	- - - -
Other Constituents	Clay, silica, carbonates	Primarily--clays, micas, carbonaceous material Often--quartz, pyrite, hematite Occasionally--calcite, ankerite, apatite, garnet, rutile, sphene, tourmaline, and zircon Primarily--silica, aluminum, carbon Secondarily--calcium, sulfur, magnesium, sodium, iron, potassium Occasionally--manganese, phosphorous Rarely--copper, nickel, zinc	the same as fresh refuse with slag material and sulfates

Source: Consolidation Coal preparation plant data. Michael Baker, unpublished. Pennsylvania State University, unpublished. U.S. Department of the Interior, 1974.

Several properties affect the desirability of coal refuse for disposal:

1. In general, the size of coal refuse is such that much of it can be disposed by pneumatic methods without crushing, but almost all of it must be crushed for hydraulic methods of backfilling.

2. The specific gravity is greater than coal but the bulk density is such that a ton of refuse occupies approximately the same volume as the volume of the void left after a ton of coal is extracted and the roof has converged slightly (approximately 1 cubic yard).

3. The optimum moisture content for maximum compaction of coal waste is approximately 10-12 percent, essentially that of its average moisture content.

4. The permeability of compacted waste is approximately that of coarse to fine sand. Uncompacted waste has a higher permeability conducive to oxidation and leaching of minerals. This can result in water pollution and/or spontaneous combustion. The permeability of hydraulically backfilled material is less than that of pneumatically backfilled material.

5. The carbon content of the coal waste ranges up to 45 percent but most waste averages less than 10 percent and should not present a fire hazard if disposed properly. When improperly handled (i.e., not compacted and allowed to segregate by size) the potential for ignition increases particularly with waste of high sulfur content.

6. Coal waste like coal itself releases its entrained methane content over an extended period of time. The crushing of coal refuse exposes more surface area and liberates some to the methane. After disposal, the refuse will continue to evolve methane slowly over time.

Changes in Properties of Coal Refuse on Aging and Weathering

Current surface waste disposal practices that conform with stringent state or federal regulations result in a relatively homogeneous pile with low ignition potential. However, when the refuse was dumped from an overhead tramway, the coarse materials rolled down to the lower part of the refuse pile and left the smaller material on the top of the pile. Such segregation enhances the likelihood of combustion and promotes its spread once it is underway. Upon long-term weathering or after further manipulation of the pile, the finer spoil can fill many of the voids in the coarse material and lessen the potential for ignition. The waste contained in the upper portions of a refuse pile and on the outside of the pile weathers and becomes finer than fresh refuse as well as finer than the refuse in the core of the old pile.

The content of coal refuse is relatively unaffected by freezing and thawing but is significantly altered by precipitation. Percolating water leaches out the products of the oxidation of iron sulfide or carbonaceous material. In the absence of adequate compaction and soil cover, the leachate products will include, in addition to iron and aluminum salts, calcium and magnesium sulfates and possibly magnesium, barium and sodium chlorides. Adequate compaction, soil covering, and revegetation of a refuse pile can result in decreased permeability to diminish the chemical, physical and biological degradation and oxidation. The permeability of backfilled waste will influence the amount of leaching of the waste material and also the ease with which associated strata are exposed to further leaching.

The composition of a refuse pile, its size, mineralogy, and chemical properties become segregated as materials are washed from high on the waste pile to the toe of the pile. Thus, weathered refuse is more likely to contain soluble sulfates than fresh coal waste. Burnt coal refuse generally contains a high percentage of sulfates. These may swell after compaction and may exhibit a greater resistance to frost heave than the original coal refuse. Weathered unburned compacted refuse will have nearly the same swelling properties as fresh refuse and vary primarily with the shale and pyrite content (Glover, 1973).

2.3 PROBLEMS RESULTING FROM PAST AND CURRENT SURFACE DISPOSAL PRACTICES

Mandatory federal standards recently proposed under the authority of the Coal Mine Health and Safety Act of 1969 would require the coal mine operator to dispose of wastes in a stable manner in a location that will not impede surface drainage, to layer and compact all refuse, and seal the surface of the pile (see Chapter 5). Well-built coal waste piles conforming to the strictest state and federal standards, constructed on relatively dry sites in thin layers, and properly compacted and sealed should not create air or water pollution problems. However, many current disposal practices do not conform to these strict standards. A 1973 survey by the Department of Environmental Resources of Pennsylvania of 135 active refuse areas revealed that 93 contribute to water pollution by acid drainage or siltation, to air pollution by burning, or exhibit signs of instability. Thirteen refuse areas showed fair to good evidence of revegetation or restoration (Pennsylvania Department of Natural Resources, unpublished).

The adverse effects of past disposal practices are glaring and include air pollution, water pollution, safety hazards, and sociological and psychological impacts.

Air Pollution from Coal Refuse Piles

Burning refuse banks were once considered one of the inevitable consequences of coal mining (Figures 2.6 and 2.7). The putrid smell, the smoke, the noxious gases and dust typical of burning refuse piles unfortunately are not uncommon in Appalachia. The adverse impact of burning waste on land values and economic development has long been recognized, but the associated physiological or psychological effects of these noxious, poisonous fumes upon human, animal, or plant communities have never been completely analyzed. It is known that emissions from burning and smouldering waste include carbon monoxide, carbon dioxide, hydrogen sulfide, sulfur dioxide, and ammonia. Gases from burning piles defoliate trees, cause crop damage, and discolor paint miles downwind. During temperature inversions, light rainstorms and drizzles, smoke and smog from such piles have created hazardous driving conditions. Most serious of all, these emissions can exacerbate problems for people with unhealthy respiratory systems such as those with chronic bronchitis, asthma, or pneumoconiosis. The extent of such conditions, the damage that has been caused or the number of lives shortened or lost has not been calculated.

Coal refuse piles may be ignited through a variety of circumstances including spontaneous combustion, careless burning of trash on or near the piles, forest fires, campfires left burning, or intentional ignition to create residues ("red-dog") which may be used for road base materials. Spontaneous combustion requires voids within the pile for moisture, circulation of air, oxidation of pyrite sufficient to create internal temperatures of at least 450° F, low ignition temperature material (timber) and finally the

FIGURE 2.6 Glen Lyon refuse bank. Looking northeast, community of Glen Lyon and #6 Breaker. Susquehanna Collieries Company on right. Photo taken April 19, 1970. The light streaks on the refuse pile left of center are flames and smoke from the burning refuse (Bureau of Mines).

FIGURE 2.7 Infrared photograph (Hg:Cd:Te N 1200' AEL) taken of Laurel Run, Pennsylvania, February 20, 1973. The light areas are fires or hot spots on refuse piles or in underground mine workings (H.R.B. Singer).

oxidation of coal. With the growth of population centers in the coal fields, urban trash and garbage are frequently disposed on coal waste piles. Since garbage dumps are traditionally ignited to reduce bulk, the consequence can be ignition of coal waste piles.

A 1963 survey indicated that of the more than 2,000 large refuse piles in the United States, more than 495 refuse piles were on fire in 15 states; 92 percent of these were located in five states (West Virginia, Pennsylvania, Kentucky, Virginia, and Ohio). Burning banks located at abandoned mines accounted for 58 percent of the total. A 1968 survey indicated that the number of large burning refuse banks in the Appalachian region had been reduced to 292, of which 74 were in Pennsylvania (U.S. Department of the Interior, 1971). Of the 135 operating refuse areas in Pennsylvania surveyed in 1972, 19 were in some state of combustion (Pennsylvania Department of Environmental Resources, unpublished). The U.S. Public Health Service, the Environmental Protection Agency, the U.S. Bureau of Mines, and several regional and state programs have sponsored demonstration projects to develop techniques for extinguishing burning refuse banks. The costs of the fire fighting techniques range from 30 cents/cubic yard of waste to several dollars/cubic yard and average approximately 1 dollar/cubic yard of waste.

External sources of ignition can be minimized through a regular inspection program. However, most states are not sufficiently staffed to routinely inspect even active waste piles for evidence of hot spots or actual refuse ignition. As a result, regulatory authorities usually learn of ignited inactive piles through public complaints long after the fire is of major magnitude.

Proper disposal procedures significantly reduce the probability of ignition of a refuse pile. The Pennsylvania Department of Environmental Resources reports that no piles or portions of piles created after 1962 (when regulations requiring layering and compaction went into effect) which were properly operated have ignited. Kentucky has had less success in preventing ignition due to inadequate regulation of refuse deposition and insufficient staff for enforcement. The state air pollution control agency is now in the process of promulgating regulations to eliminate this situation.

Water Pollution Problems from Coal Waste Piles

Two general sources of water pollution can result from waste piles: physical pollution such as siltation, and chemical pollution such as acid drainage. In the former, fine material is carried from the waste pile into streams by rain water runoff causing increased siltation downstream. Most states require that water containing suspended solids from coal waste banks be collected and treated in the same manner as the "blackwater" or processing water produced by coal preparation plants. However, the large-scale financial impact of inappropriate surface disposal has been analyzed in a recent GAO report on the impact of coal mining on federal reservoirs. The useful life of such dams for flood control has been reduced by accelerated siltation primarily from surface mining operations but also from waste disposal areas (U.S. Comptroller General, 1973).

A far more serious and complex problem is chemical water pollution. In high-sulfur coal and coal waste there is significant potential that the sulfur will react chemically with air and water to produce sulfuric acid unless proper disposal techniques are used. Other ions including iron, aluminum and manganese are produced as part of the reaction creating acid water. The effect of this on streams has been to lower the pH below tolerance level of many desirable aquatic life species. The heavy metal constituents are toxic singly and may act synergistically, eliminating less resistant plant and animal species. Ferric hydroxide, "yellow boy," a precipitate resulting from acid drainage, smothers lifeforms and coats stream bottoms, thus reducing percolation and oxygenation of the water, and limiting the available breeding areas for aquatic species.

Conscientious compaction of coal refuse, construction of water diversion ditches, soil covering and successful revegetation of coal refuse piles can essentially eliminate water pollution after the use of the pile is terminated. However, these piles must be constructed with diligence and with strict observation by enforcement agencies because once a source of water pollution is created, it is extremely difficult to correct. According to a report of the U.S. Department of the Interior, over 2,000 mineral waste banks contribute to stream pollution in the United States (U.S. Department of the Interior, unpublished, a). A typical coal refuse pile may produce on the average of 1.5-2.0 pounds of acid per acre per day and 0.5-0.7 pounds of iron per acre per day. Acid production as high as 305 pounds per acre per day was reported at one site in Illinois (U.S. Environmental Protection Agency, 1971).

Of the 135 active waste piles examined by the Department of Environmental Resources in Pennsylvania in 1972, 61 contributed to acid pollution and 60 contributed to stream siltation. Sixty-nine of the 135 piles surveyed are within 1/4 mile of stream banks. The percentage of abandoned piles with acid problems is probably higher. The effects of physical and chemical water pollution from past and present mining activities have contaminated 10,500 miles of streams in Appalachia (Appalachian Regional Commission, 1969). Approximately 7.5 percent of the acid pollution is contributed by coal processing plants and refuse areas. Storm runoff and the placement of piles near streams increases the pollution effects.

In order for the states ultimately to meet the water quality criteria approved by the EPA, polluting water runoff from coal refuse piles will have to be abated. In Pennsylvania, recent regulations require treatment of such drainage as a condition for receiving a coal refuse permit. Thus, the acid problems of runoff water from active refuse banks in Pennsylvania may be significantly improved in the near future.

Both surface disposal and underground disposal of coal mine waste put potentially polluting material where it can result in chemical water pollution. Acid drainage from inactive underground mines is the most difficult source of acid drainage to control and the greatest single source of mine drainage pollution. It accounts for 70 percent of acid drainage from all surface and underground operations (Appalachian Regional Commission, 1969). Backfilling would retard the free flow of water in an inactive mine, and if compacted by the overlying strata it would further reduce percolation through the refuse itself. This could result in partial inundation of the mine thereby reducing acid water generation. In returning mine waste underground by hydraulic methods to flooded, abandoned workings, the difficulties of controlling acid drainage may be aggravated if the area is already a source of acid drainage. However, the disposal of waste in permanently submerged portions of the mine would not significantly or permanently worsen water quality. Thus, the effect of backfilling might be to alleviate or minimize acid drainage problems in recently mined or in active mine operations, but might exacerbate acid conditions in extant acid mine pools.

Safety of the Surface Environment

Historically, little attention has been given to coal waste disposal. Little or no predisposal site preparation or impoundment design was considered necessary. The result was the creation of many, large, undesigned and poorly constructed coal waste piles and impoundments. Many of these constitute a safety hazard. The magnitude of these hazards was brought home to the public after two major disasters: a flowslide of 140,000 cubic yards of waste from a 200-foot waste pile at Aberfan, South Wales in 1966, and a failure of a waste impoundment at Middle Fork, Buffalo Creek, West Virginia in 1972 when 650,000 cubic yards of water and 220,000 cubic yards of material were transported downstream. The cause of both slides was improper disposal of coal mine waste. In each case the physical cause was saturation of the waste by water which reduced its stability to such an extent that the material flowed as a liquid. The geologic cause of the Aberfan disaster was the disposal of waste over a spring of water. The climatic cause of the Buffalo Creek disaster was a rainstorm which deposited 3.7 inches of rain in 72 hours (a 2-3 year frequency storm). Both incidents were avoidable and inexcusable. A description of the events taken from the official reports follows.

Aberfan, South Wales, October 21, 1966:

At about 9:15 a.m. on Friday, October 21st, 1966, many thousands of tons of colliery rubbish swept swiftly and with a jet-like roar down the side of the Merthyr Mountain which forms the western flank of the coal-mining village of Aberfan. This massive breakaway from a vast tip (pile) overwhelmed in its course the two Hafod-Tanglwys-Uchaf farm cottages on the mountainside and killed three occupants. It crossed the disused canal and surmounted the railway embankment. It engulfed and destroyed a school and eighteen houses and damaged another school and other dwellings in the village before its onward flow substantially ceased.... despite desperate and heroically sustained efforts of (the many people of) all ages and occupations who rushed to Aberfan from far and wide, after 11 a.m. on that fateful day nobody buried by the slide was rescued alive. In the disaster no less than 144 men, women, and children were killed. Most of them were between the ages of 7 and 10, 109 of them perishing inside the Pantglas Junior School. Of the 28 adults who died, 5 were teachers in the school. In addition, 29 children and 6 adults were injured, some of them seriously. 16 houses were damaged by sludge, 60 houses had to be evacuated, others were unavoidably damaged in the course of the rescue operations, and a number of motor cars were crushed by the initial fall. According to Professor Bishop, in the final slip some 140,000 cubic yards of rubbish were deposited on the lower slopes of the mountainside and in the village of Aberfan. (Report of the Tribunal appointed to inquire into the Disaster at Aberfan, 1967, p. 26)

Buffalo Creek, West Virginia, February 26, 1972 (see Figures 2.8 and 2.9):

Approximately 21 million cubic feet of water was released from the coal-refuse dams on Middle Fork (Saunders, Logan County, West Virginia) beginning at about 8:00 a.m. on February 26.... The previously impounded water then began its wild 17-mile plunge down Buffalo Creek falling more than 700 feet in its race from Saunders to Man.... All homes and structures at Saunders were totally destroyed.... The flood wave traveled from Saunders to Pardee in about 10 minutes at an average velocity of 19 feet per second.... The flood waters arrived at Lorado at about 8:15 a.m. The flood flow was 6 to 8 feet deep on the flood plain and almost completely destroyed the town. A few well-constructed buildings survived, but nearly all homes of wooden construction erected on a slab foundation were demolished.... Flood damage downstream from Amherstdale, although still serious, was far less extensive.

The flooding resulted in the confirmed deaths of 116 persons as of the date of this report (March 12, 1972), total destruction of 502 permanent home structures and 44 mobile homes, major damage to 268 additional permanent home structures and 42 mobile homes, and minor damage to 270 additional homes along Buffalo Creek from Saunders to Man, West Virginia, a distance of about 17 miles. It was estimated that about 4,000 persons were left homeless. Numerous homes in the Buffalo Creek area that were located above the flood plain escaped damage.

The flooding also destroyed about 1,000 automobiles and trucks, several highway and railway bridges, sections of railroad tracks and highway, public utility power cables and poles, telephone lines and poles, and other installations. Mine refuse silt, and debris were scattered for miles along Buffalo Creek. About 30 persons who resided in the Buffalo Creek area remained on the missing list. (U.S. Department of the Interior, 1972, pp. 17-22)

Other horror stories have been associated with waste piles: children have been periodically burned, and minor slides have claimed lives. Twelve known fatalities have resulted from explosions on burning or smouldering waste piles. The accidents are not a thing of the past. The most recent occurrence was a multiple fatality January 30, 1974, in West Virginia where two bulldozer operators were killed while modifying a smouldering pile

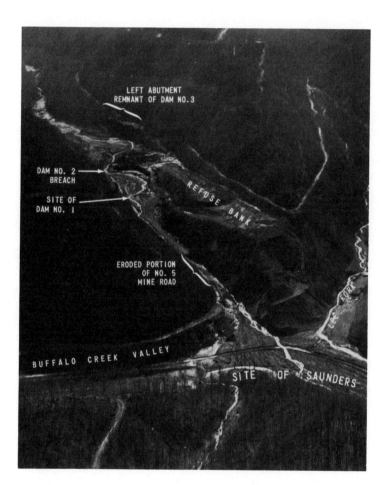

FIGURE 2.8 Middle Fork Valley, February 28, 1972. (Photograph courtesy West Virginia Department of Highways)

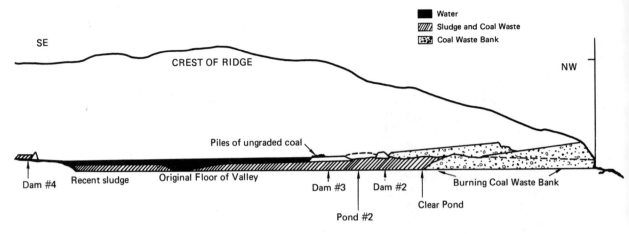

FIGURE 2.9 Cross section of the Buffalo Creek waste bank and impoundments before failure. (Bishop, 1973, p. 373 after Davies, 1972.)

(United Mine Workers of America, 1974). After Hurricane Agnes in 1973, five waste impoundments gave way in Pennsylvania alone. Only one of these impoundments had been considered an imminent hazard in a 1972 survey of impoundments. In 1972, an impoundment burst through auger holes into sections of a mine in eastern Kentucky. Fortuitously, the incident occurred on a Sunday and no one was killed (Mountain Eagle, 1972). Over 33 percent of the impoundments surveyed after the Buffalo Creek disaster were considered potential hazards (U.S. Department of the Interior, 1972c, 1973d).

These unnecessary hazards to a community's security have been brought about by some unthinking coal operators working under indifferent and irresponsible enforcement of poor legislation. Current good practices of waste-disposal will not result in a dangerous environment for those who live or work near the pile. Nevertheless, many inactive or abandoned impoundments and piles are unstable or contribute to air or water pollution. These piles and impoundments continue to present hazards to the safety and health of the public.

Problems of Subsidence

A strong potential for subsidence exists where a significant amount of coal has been extracted (refer to Appendix B for an extensive discussion of the effects of underground mining upon overlying strata and upon the ground surface). Underground mining for coal has been responsible for 95 percent of the mine subsidence in the United States. This subsidence over coal mines has affected approximately 2 million acres of which approximately 150,000 acres have been in urban areas (U.S. Department of the Interior, unpublished a). Occupants of surface areas affected by subsidence have experienced

FIGURE 2.10 Subsidence-damaged house over an abandoned coal mine, Coaldale, Pennsylvania.

buckling of streets and sidewalks. Homes and other structures have been severely damaged, and utility lines have been broken (Figure 2.10). Broken gas lines have proved to be extremely dangerous and in many cases residents have been forced to abandon their homes.

Subsidence-prone areas can be affected by either localized and random sink hole occurrences or by general surface subsidence movement depending on the geology, mining method, and the condition of the remaining coal pillars (Figures 2.11 and 2.12). Differential horizontal components of strain can seriously damage surface structures in either random or general subsidence. Where areas settle uniformly damage can be minor. Backfilling can prevent subsidence by lending lateral support to pillars and can reduce subsidence by reducing the volume of the void created by mining. Maximum subsidence is reduced approximately one-third by total backfilling in areas of complete extraction.

Psychological and Sociological Effects of Past Disposal Practices

Most travelers in the coal regions of the United States and Europe realize at once that they are in a mining area from the sight of the ever present grey-black coal waste piles. Although local residents are said to grow accustomed to such piles and generations have been raised with waste piles as the backdrop for their homes, the presence of these piles is generally not considered aesthetically acceptable and the tolerance of the local population for them appears to diminish as the economic dependence of local citizens on the coal industry diminishes. The piles act as serious hindrances for the chambers of commerce and other industrial development groups trying to lure new industry to coal-depleted areas. Proper surface disposal methods can disguise the waste piles so they blend into the surrounding area. Underground disposal would remove such waste from sight.

Surface disposal of coal waste also creates conflicts of land use. It is estimated that the coal waste piles and impoundments accumulated between

FIGURE 2.11 Marshall, Colorado, 1973. Collapse of rock into underground workings outlines rooms and pillars of an abandoned coal mine (most visible in the lower center of the photograph). The fracture planes serve as surfaces for slippage and subsidence. Collapse is continuing even though mining ended about 40 years ago. (U.S. Geological Survey)

FIGURE 2.12 Marshall, Colorado, 1973. Collapse of rock into underground workings outlines rooms and pillars of an abandoned coal mine 1/2 km northeast of Marshall, Colorado. Above the partially mined area, the water in the irrigation ditch must be confined within a metal flume to prevent loss through cracks in the rock into the underground workings. Periodic repairs are required due to continuing subsidence due to current burning of coal in this area. (U.S. Geological Survey)

1930-1971 cover approximately 225,000 acres of land (U.S. Department of the Interior, unpublished b). Surface disposal is not encompassed within local jurisdictional or zoning policies; and seldom are future land uses of the completed site considered.

In many areas the coal preparation plants and the coal refuse piles are located in valley bottoms adjacent to the transportation network of roads and railroads. At these locations the refuse piles occupy land suitable for development since flat sites near transportation routes are a scarce commodity in mountainous terrain. Foresight in selecting a disposal method could avoid a permanent allocation of this valuable flat bottomland to coal waste disposal and permit its use for development.

The threat that subsidence hazards hold for present and future urban areas has grown progressively more serious as cities grow in areas overlying abandoned mines with deteriorating mine pillars. The threat of disasters such as Buffalo Creek continued to haunt those who survived and those who live downstream from similar embankments (Lifton, 1973). Much has been written about the social costs of mining coal as evidenced by the Appalachian region and its people. It has been argued that past disposal practices have resulted in rural blights, which, like their urban counterparts, have had a negative impact on the lives of the people who inhabit the region. The overall effect is impossible to measure. Environmentally acceptable surface disposal methods need not mar the beauty of the landscape, pollute the air and water, or endanger public safety. Underground disposal of mine wastes would circumvent these problems. However, the alternatives for current or future disposal do not rectify the problems inherited from the past in the form of inactive and abandoned disposal areas.

2.4   ALTERNATIVES TO DISPOSAL OF UNDERGROUND COAL MINE WASTES

Although the charge to this committee is solely to evaluate the feasibility of returning coal mine wastes underground the committee could not fail to note that both the coal refuse and the mine void may be resources in themselves.

Although little effort has been made to market coal mine waste, some alternative uses are possible. The quantity of waste that could be so utilized and competitively marketed without incentives is difficult to determine. Most uses would consume a relatively small amount of annual coal waste production. Bulk use such as construction fill is restricted by transportation, material characteristic, design and engineering factors. However, local shortages of concrete aggregate and construction fill in some urban areas and highway projects could provide a form of useful disposal of abandoned or active mine wastes. Coal mine waste is an asset, but it is generally not located where it can be most useful.

Some old waste piles consist of material with high enough carbon content to be recycled for their fuel content and burned without reprocessing in pulverized fuel stokers. Reprocessing of such wastes for coal allows reconstruction of waste piles in an acceptable fashion. Alternate uses of coal mine waste include: construction fill, manufacturing lightweight aggregate, road base, anti-skid material for roads, brick manufacture and materials for insulation (Table 2.2).

Construction Fill

Coal mine wastes are used as construction fill in highway embankments and sites where there is a demand for fill materials. In general it has proved economical only where it has been subsidized or where the cost of excavating fill has been greater than the cost of hauling coal mine waste. As with disposition of all fill, good engineering practice requires placement and compaction in thin layers.

TABLE 2.2 Tabulation of Alternate Uses of Coal Mine Waste and Their Proven or Estimated Feasibility

Proposed Use	Amount of Coal Refuse Used	Experience Abroad	Experience in the United States	Economic Feasibility
Recycled for fuel	A significant amount very old waste piles	Yes	Yes	Dependent on percentage carbon content
Construction fill	As much as 10% annual coal waste production	Yes	Yes	Locally good
Lightweight aggregate	Not much	Yes	Yes	Not usually
Road base and anti-skid materials	Some	Yes	Yes	No
Brick making	Very little	Yes	Experimentally	No
Others: mineral wool recycled for noncarbon constituents	Very little	No	No	No

In mining areas where there is a shortage of relatively flat land suitable for development, coal mine waste could be used as fill to construct more property for development. In areas where both underground and surface mining are present, the coal waste from underground mines could be backfilled in surface mined areas.

Coal mine waste has been used as fill in the United States, Britain, and Germany for airfields, industrial sites, housing developments, and road embankments. Highway systems in areas of waste piles can be located and designed to utilize as much coal waste as possible. Even so, these construction uses of waste in Britain consume less than one-tenth the annual production of coal waste (National Coal Board; Bishop, 1973).

The use of coal mine waste for dike construction in Germany was reported as early as 1926. Coal waste from England may be transported to the Netherlands for dike construction. Land fill with coal wastes in the Netherlands has provided land for agriculture and other land development.

Lightweight Aggregate, Road Materials

Coal mine wastes generally require special processing for other than construction purposes. In areas of limited supplies of sand, gravel or rock, coal mine wastes may be suitable after crushing and treatment for manufacture of concrete aggregate or lightweight aggregate. Lightweight aggregates are produced from clay, shale, slate or mine waste by heating processes but the best quality aggregate requires select materials which will expand. For this particular use, fresh refuse will be more suited than weathered refuse because it expands more completely. Lightweight aggregate made from coal waste has been used in Poland for building precast wall panels.

Aggregate for use in asphaltic highway surfacing and patching has been made from crushed and screened burned coal waste both in Europe and in the United States at Dante, Virginia. Coal mine waste mixed with 5 to 10 percent cement has been used as a road base and could be used extensively should other granular base materials become scarce. Experimental work in Europe appears promising for a skid-resistant highway topping made from a mixture of bauxite and burned coal mine waste.

Brick making from coal mine waste has been the subject of research in this country and has been utilized in the Netherlands and in England. Advantages include: (1) the carbon in the waste results in lower fuel consumption, and (2) the burned-out carbon results in lightweight brick. However, manufacture of such bricks has not proven to be economically competitive with clay bricks.

Other possible uses of coal mine waste include the manufacture of mineral wool as insulating material and the extraction of aluminum or other elements from the waste. However, most of the 46 elements that have been found in coal mine waste are present in such small concentrations that costs are too great to justify their extraction.

2.5 POTENTIAL AMOUNT OF MATERIAL USE

At this time, the only prospect for utilizing large quantities of coal mine waste appears to be as construction fill. Use of coal waste for light-weight aggregate, brick, and for anti-skid material may be promising, but not in sufficient quantity to solve the waste disposal problems.

Germany in 1966 and Britain in 1968 set up formal programs to encourage the use of coal mine waste. The British National Coal Board analyzed the properties of waste piles in their ownership, made a thorough inventory of possible market uses of coal waste and provided technical advice to potential users of coal waste. However, even during a large-scale road building program only 6 percent of the country's yearly production of underground coal mine wastes were used for road fill (British Department of the Environment, 1972). A similar technical and marketing effort in the United States could help the coal industry provide useful construction materials to potential users. However, the problem of disposing of the bulk of the annual production of underground coal mine wastes will remain as well as the problem of the past accumulation of waste.

2.6 ALTERNATIVE USE OF UNDERGROUND SPACE

Underground space has many uses. Limestone mines in Kansas and Pennsylvania are used for office space and storage. In Kansas, large warehouse facilities have been developed in abandoned limestone mines, and new mines are being developed to facilitate future use. In Sweden, large underground areas have been excavated for similar purposes. Coal mines have not been utilized in this manner for all or some of the following reasons: (1) generally, mined seams have low roofs and lack cavernous chambers; (2) roof collapse in coal mines is more likely than in limestone or hard rock mines; (3) the danger of gas continues after mining ceases; (4) acidic groundwater could affect construction materials; (5) complete extraction of pillars eliminates future access to the voids. The number of old, shallow mines suitable for commercial use appears to be small. Local schemes to use abandoned coal mines for growing mushrooms, storage places for junked automobiles, or for disposal of wastes other than coal wastes have met with varying degrees of success. Such uses do not promise to become widespread. The trend to deeper mines utilizing longwall mining to extract a large percentage of the coal indicates few desirable long-term, safe underground openings will be developed.

CHAPTER 3
TECHNICAL FEASIBILITY OF UNDERGROUND DISPOSAL OF WASTES

SUMMARY

Underground disposal, or "backfilling," of coal mine waste has been clearly demonstrated to be technically feasible in active longwall mining operations in Europe and in abandoned room and pillar mines in the United States. However, the practice has declined drastically in Europe, being considered a drawback in mechanized, high production-rate mining; although a recently developed lateral discharge pneumatic backfilling system appears to overcome some of the earlier disadvantages.

Most U.S. mines use the "room and pillar" mining system, but the "long-wall" mining system, for which present backfilling methods were designed, is becoming more widely adopted in the drive to obtain higher production but currently accounts for less than 5 percent of underground coal mine production.

A safe, operational system of concurrent backfilling and extraction at operating room and pillar faces has not yet been developed, and would require substantial development efforts and operational changes.

Backfilling of abandoned mines or abandoned sections of operating room and pillar mines through boreholes from the surface appears technically less difficult than concurrent backfilling and extraction in active sections of the mine. Such systems could involve some surface neighborhood inconveniences and also would be limited to those situations where fill transportation distances from the preparation plant to the mined out areas are not excessive.

The health and safety of underground miners could be affected by underground disposal systems, particularly those where concurrent mining and backfilling take place in close proximity. In particular, the increased congestion at the face, the increased dust and noise of pneumatic systems, and, in hydraulic systems, the hazard of barricade ruptures, would present safety hazards.

In general the economic costs of underground disposal exceed *by far* the normal costs of surface disposal. Only if the physical conditions at a mine site for environmentally acceptable surface disposal were so adverse or the social benefits of underground disposal so great would underground disposal become economically competitive with surface disposal for modern, high production rate coal mines.

3.1 CURRENT METHODS OF UNDERGROUND COAL MINING

For the purposes of this discussion of backfilling for disposal, it is sufficient to discuss the three principal methods of mining coal:

1. Room and pillar mining--The traditional method in the United States, this method currently accounts for over 90 percent of all domestic underground coal production.

2. Longwall mining--This is the principal method used in the coal mines of Europe. Although longwall mining now accounts for less than 5 percent of U.S. underground coal production, it is increasing in popularity, largely because of the more continuous nature of the mining operations, the greater percentage recovery, and because recent developments in mechanized longwall equipment have improved its efficiency.

3. Shortwall mining--Very little used at present, this method is attracting attention in the United States because it combines some of the advantages of room and pillar mining and of longwall mining.

There are many variations on these principal methods and there are other systems specially devised for unusual geological conditions. For example the Atlas of Mining Methods, compiled by an international group of mining experts, lists 56 underground mining methods for extraction of coal deposits (UNESCO, 1963).

Room and Pillar Mining

Because the removal of coal from an underground seam produces a tendency for the overlying rock to collapse, mining methods are designed to prevent accidents to personnel or damage to equipment. In room and pillar mining the coal is extracted in two main stages. In the first stage the coal is mined in a pattern of rooms separated by pillars of unmined coal (Figure 3.1). The room should be wide enough to allow effective passage of mining machinery, but narrow enough to avoid risks of roof collapse. The roof is usually reinforced by "rock bolting" and other forms of support to guard against roof falls. The width of the pillar is such that it is at least sufficient to support the overlying rock without collapse. Pillar widths usually range from 1 to 3 times the width of the room, depending on roof conditions and depth. The required size of the pillar will increase as the depth of the mine increases. Pillar dimensions are also influenced by procedures used in subsequent (stage 2) extraction operations. Ideally, the entire coal seam is developed as a regular pattern of pillars throughout the mining property. If first-stage mining only is practiced and the pillars are left in the ground, the mining method is known as "partial extraction." Over 70 percent of U.S. mines practice partial extraction in at least one section of the mine (U.S. Department of the Interior, unpublished).

The second stage, pillar extraction, begins when the boundary of the working area is reached (Figure 3.2). Coal is extracted from the pillars

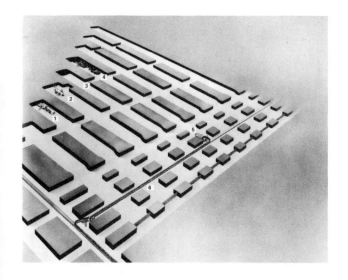

FIGURE 3.1 Conventional room and pillar mining method. The numbered areas indicate six operations: 1) undercutting of the coal seam; 2) drilling holes for blasting; 3) blasted coal; 4) loading machine and shuttle car; 5) shuttle car discharging coal onto a conveyor belt; 6) conveyor belt carrying the material out of the mine. (Joy Manufacturing Company)

FIGURE 3.2 Continuous room and pillar mining method showing a development section (stage 1) and a pillar retreat section (stage 2). Both sections are being mined by "continuous" mine machines rather than by "conventional" mining machines in Figure 3.1. A conveyor belt transports coal mined in the development section to the trunk conveyor. The trunk conveyor carries the coal out of the mine. It is located in one of 5 entries forming the main heading. (Joy Manufacturing Company)

while "retreating" toward the main entry. Successive cuts are taken to remove the pillar while the miner is protected by roof support from posts, cribs, and bolted roof (Figure 3.3). Elimination of these pillars removes support from the overlying rock which consequently collapses or "caves" into the mined out area after the miners have moved to the next pillar.

Adopting a stepped retreat line allows pillar extraction to be carried out under relatively safe conditions where the roof collapse does not occur within the working area. Some danger does exist, however, and it is usually necessary to leave some portions ("remnants") of the pillars for worker and machine protection during extraction operations. (Note the shaded areas in

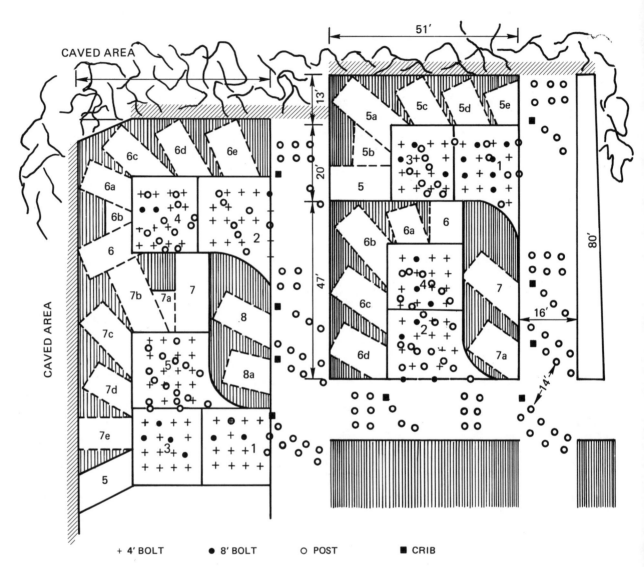

+ 4' BOLT   ● 8' BOLT   ○ POST   ■ CRIB

FIGURE 3.3 Pillar extraction and plan for installing roof supports. The pillar to the left is removed by cuts as successively numbered in the diagram. Then the pillar to the right is removed. This system is designed to have the miners under supported roof at all times. The shaded area is not mined. (Consolidation Coal, Mathies Mine)

Figure 3.3.) These remnants crush as the extracted region is enlarged. In
this manner the coal extraction operations progressively retreat toward the
mine shaft or main entry removing as much of the remaining coal as possible.
This is known as "complete" extraction even though not all the coal is
removed.

From the standpoint of safety, pillar extraction is a particularly
difficult part of room and pillar mining, particularly where the overlying
roof is weak or the pressure due to the weight of overlying rock is high.
Under such circumstances, a sizable fraction of the seam is left unmined.
In addition, some areas of the coal seam must be left unmined in order to
protect major access and egress haulageways and also to protect important
surface structures from damage due to subsidence. The average percentage
extraction in room and pillar mining in U.S. mines currently is approximately
58 percent. This percentage decreases with the depth of mining and the method
becomes progressively less economic.

Some important characteristics of room and pillar mining are:

1. It allows coal seams to be partially extracted with minimum disturbance
of the overlying rock strata and without surface subsidence, provided adequately sized pillars are left permanently and do not deteriorate. When
pillars deteriorate, delayed collapse can occur and produce surface damage
many years after mining has been completed (Figures 2.11, 2.12 and Table 2.2,
Chapter 2). When pillars are extracted ("complete extraction") surface subsidence will occur.
2. The pillars act as roof supports and allow mining to be carried out with
a minimum of "artificial" supports (timber, steel, etc.). Usually roof bolting
to prevent fall of the immediate roof is all that is necessary with partial
extraction.
3. Special attention must be given to the problem of providing adequate
ventilation to the areas being extracted.
4. Mining is achieved with mechanized, mobile equipment. The high degree
of mechanization, the efficient methods of roof support and the system's inherent flexibility have led to the highest average productivity in the world.

Longwall Mining

The room and pillar system is the method that has been almost exclusively
used in shallow mines, but it becomes progressively less economic as mines
become deeper. In Europe, with its geological conditions and prevailing
state of mining technology and economics, a major switch from room and pillar
to longwall mining took place approximately 30-40 years ago, as the shallow
coal seams were mined out and most new coal mines were developed 500 m-600 m
(1,500-2,000 feet) or more in depth. The longwall method is now almost exclusively used in underground mines in Britain, Germany and Belgium, and in
the majority of mines in France and Poland (5th International Strata Control
Conference, 1972).

The potential of the longwall system for high productivity in appropriate
geological conditions has recently attracted considerable interest in the

United States for coal mines at depths that average substantially less than those in Europe. However, in 1972, longwall mining accounted for only 2.5 percent of U.S. underground production (U.S. Department of the Interior, 1973d).

In the longwall method, most of the seam is extracted by more or less continuously slicing off the coal along a series of long faces that are of the order of 300 or more feet in length. In longwall advance mining, the coal seam is extracted as completely as possible, progressing from the main entry toward the boundaries of the property. In longwall retreat mining, the haulageways necessary to provide access, service and ventilation are first driven (to the boundaries of the property or some other limit), and then faces are developed to extract the coal by retreating toward the shaft (Figure 3.4). The advance methods allow a mine to reach full production operation soon after shafts are developed to the coal seam, but tend to result in difficult roof control and haulageway maintenance costs. Conversely, retreat systems involve longer delays in start-up and return on capital investment, but result in better working conditions, since roof collapse and associated ground disturbances are "left behind" the face, i.e., occur outside the active areas of the mine. Mines also are developed using a combination of advance faces and retreat faces.

The longwall method allows a greater proportion of the coal seam to be extracted from deeper deposits than is possible with the conventional room and pillar method. The frequent reference to the longwall method as "full extraction" systems can, however, be misleading. Although full extraction of the longwall panels is realized, it is still necessary to leave areas of the coal intact to protect major haulageways, and to avoid ground disturbance and surface subsidence in critical areas (see Appendix B). The overall extraction of coal in longwall mines is usually less than 80-85 percent of the seam. In extracting this high proportion of the seam, it is inevitable that

FIGURE 3.4 U.S. longwall retreat (with multiple entries). The cutting machine shown in this example is a rotary shearer. Roof supports protecting the men and machines are hydraulically controlled and advance as the longwall mining machine advances. A conveyor belt traveling below the longwall cutting machine transports the coal to the main conveyor which carries it from the mine. (Joy Manufacturing Company)

the overlying rock will collapse into the excavated space. Special supports that advance with the face are provided to maintain collapse sufficiently far behind the face to protect miners and machinery from the dangers of a collapse (see Figure 3.5 and a detailed diagram Figure 3.11). The magnitude of the surface subsidence will depend on the thickness and aerial extent of the mined out area, and the extent to which the voids are filled or supported in any way.

The U.S. longwall method differs somewhat from the European method. Federal mine safety regulations require that the haulageways for initial development of longwall panels be driven as a series of multiple entries in the seam. During extraction of the longwall panels the entries are left intact to provide ventilation paths for the draining of methane from the working face, this collapsed material (gob) and the roof above the mined out area.

Shortwall Mining

The shortwall method is a variation of the longwall method in which conventional continuous mining machines used in room and pillar mining are used to remove coal along a 100-200 foot panel face (Figure 3.5). As with room and pillar mining, shuttle cars and conveyors are used for haulage. The men and mining equipment are separated from the caved area by a row of self-advancing hydraulic roof supports. The use of continuous mining equipment enables the face to be mined in 10- or 12-foot lifts (cuts along the base) which is several times that obtained by the cutting machines used in longwall operations.

3.2 METHODS OF BACKFILLING AND APPLICABILITY

3.2.1 Methods

Pneumatic Backfilling: Standard, Older System

The classical (European) method of pneumatic backfilling on a longwall coal face is shown diagrammatically in Figure 3.6. The waste is transported along the tailgate haulageway by rail or conveyor belt and introduced into the stower where it is entrained in a flow of high pressure air. It is projected along the tailgate in special abrasion-resistant pipes, forced around a bend in order to travel parallel to the face, and ejected immediately behind the working face into the area that has been mined out. A "brattice cloth" (heavy sacking material) progressively extended along a row of wooden props, is erected to contain the waste within the backfilled area. Backfilling is interrupted as each "pipe-length" of backfilling is completed, and a section of pipe (2-3 m, 6-9 ft) is uncoupled and placed in position for

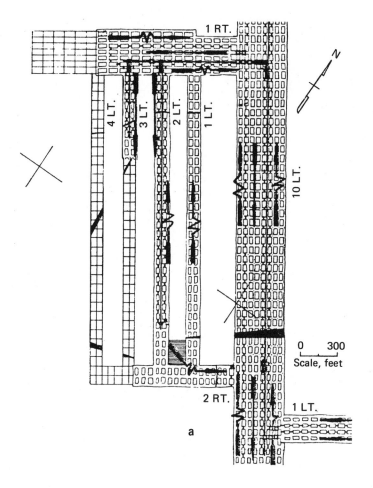

FIGURE 3.5a Shortwall mining. Shortwall area of the Hendrix No. 22 mine. Panels are 2400 ft long and contain about 65,000 tons of coal.

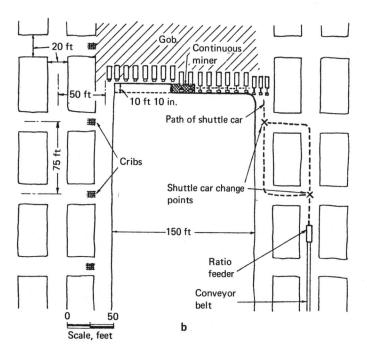

FIGURE 3.5b Shortwall arrangement at Hendrix No. 22 mine. Mining is done in such a way as to keep miner operator under supports at all times. (Palowitch, 1973)

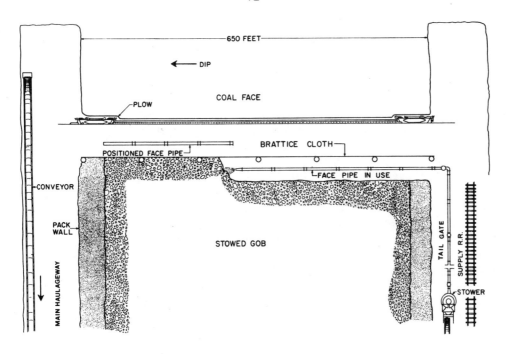

FIGURE 3.6 Conventional pneumatic stowing layout for a longwall face. (Singh and Courtney)

the next stowing sequence. Then stowing resumes. This cycle is repeated until the backfill has been placed along the entire length of the working face. Then the coal is mined to a depth of 1-2 m (3-6 ft), conveyed along the face, discharged onto a conveyor belt in the main haulageways, and transported out of the mine. Pack walls, constructed by hand of large waste rock broken down from the roof to increase head room in the main haulageways help support the roof above the main haulageways and serve to contain the backfilled waste. Waste crushed to a maximum size of one-third the pipe's inside diameter can be carried several hundreds of meters in pipes, but 150-200 m is considered the desirable maximum. Under suitable air pressure and seam height conditions waste can be projected about 50 m from the nozzle. The air pressure is up to 4 kg/cm^2 (60 psi).

This method is best suited for the three 8-hour shift daily cycle--in which coal mining is carried out on one shift, stowing and haulageways advancing on the next and general equipment maintenance on the next, or on the four 6-hour cycle in which coal mining and stowing are carried out on alternate shifts. It has not proved well suited for the continuous (24-hour) coal mining longwall method in use today in Europe or in the United States.

Pneumatic Backfilling: Lateral Discharge System

The rise in labor and capital costs in European coal mining have resulted in highly mechanized mining systems and increased outputs which are dependent on rapid and continuous advance of the longwall face. This development, the shift to a 24-hour continuous mining cycle, and the relative inefficiencies and cost of backfilling led to a substantial decline in backfilling through-out the 1960's. However, more efficient stowing machine designs developed over the past several years in Germany lower the cost of handling the stowing materials and are capable of keeping pace with continuous longwall miners (Figure 3.7a). The incorporation of regularly-spaced, side-discharge deflectors into the face pipes (Figures 3.7b, c) and overall mechanization of the stowing operation have made pneumatic backfilling compatible for working concurrently on rapidly advancing mechanized longwall faces (Figure 3.8). The waste is discharged laterally into the stowage area having been deflected from the pipe via a side-discharge unit. The side-discharge unit can be rapidly opened and closed hydraulically, without interruption of the waste delivery. The stowage area is filled progressively along the face by first opening the furthermost deflector to fill the first section to be backfilled; then closing this deflector, and simultaneously opening the next. The face pipes are attached to the rear of the (self-advancing) mechanized face support system as shown (Figure 3.9) and are advanced with the face, eliminating the tedious pipe dismantling and reassembly of the traditional system (Table 3.1). The brattice is usually replaced by heavy rubber belt guards attached to the base of the rear support. Telescopic sections of pipe in the tailgate road permit the system to be advanced continuously with the longwall face.

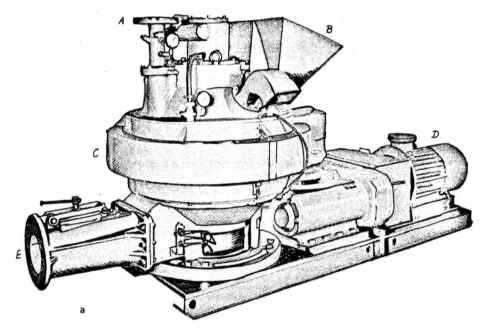

FIGURE 3.7a   Brieden Type KZS 250 Pneumatic Stower. Stowing Capacity 250 m³/hr. = 500 tons/hr.   Air under pressure is admitted at A.   Waste is dis-charged into hopper at B and is rotated into the high pressure paddle wheel C, (rotated by motor D). The waste is then projected along pipe E to the face.

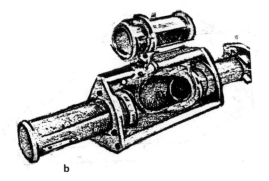

FIGURE 3.7b Hydraulic Side-Discharge Deflector. The section of 'straight-through' pipe can be quickly interchanged with the deflector shown in position in the waste stream.

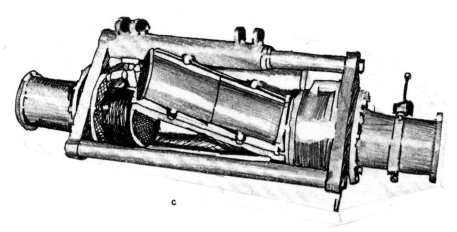

FIGURE 3.7c Mechanized Side-Discharge Unit. More recent than that shown above.

The side-discharge system can reduce the cost of pneumatic stowing to about 50 percent of that of the traditional system, but experience with it to date has been too limited to obtain reliable cost figures. Mechanized side-discharge backfilling, although still in the experimental stage, is currently being used in several European mines (Table 3.1). The committee based its estimates for the cost of pneumatically backfilling upon a side-discharge system (section 3.4 and Appendix C).

Hydraulic Backfilling

Hydraulic backfilling of waste in coal mines is not used as extensively as pneumatic methods in Germany and Poland but it is used more extensively than pneumatic methods in Belgium and France (5th International Strata Conference, 1972). Hydraulic stowing has been used in active U.S. coal mines in Sunnyside, Utah and in the Anthracite field. However, no U.S. coal mines backfill hydraulically at this time. In a typical installation in Poland crushed washery waste is placed at the surface into a large, 50-meter deep underground bunker.

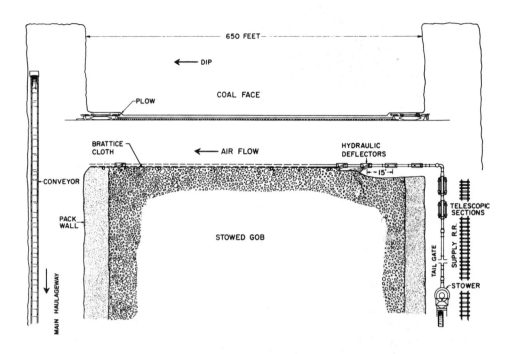

FIGURE 3.8   Lateral pneumatic stowing layout for
a longwall face.   (Singh and Courtney, 1974)

When needed, the waste is flushed from the bunker using hydraulic monitors
(high pressure water jets) and conducted several hundreds of meters down a
vertical pipe to abrasion resistant pipes.   These carry the slurry along
inclined ventilation and service haulageways to the longwall (retreat) coal
faces.   The material is backfilled at a depth of approximately 800 m using
the hydraulic pressure produced by the full hydraulic head of the 800 m high
pipe system.

Brattice cloths are erected along the face as with traditional pneumatic
stowing, and lateral discharge pipes are used to direct the waste-laden water
behind the cloth.   The removal and relocation of pipes is very similar to
that described for the traditional pneumatic system.   Stowing is carried out
up the face to allow water to drain away into a sump in the downward dipping
tailgate ahead of the face.   The water in the waste drains quickly so that
within twenty minutes of stowing the waste appears as dry as pneumatically
stowed waste.   No coal mining operations are permitted at the face during
hydraulic stowing, because of the possibility that the brattice cloth will
burst, and the water and waste would flow onto the face.   The potential
hazard of waste and slurry flowing onto the face area during hydraulic backfilling suggests that the method is less likely to be used in conjunction
with fast-moving mechanized longwall faces than pneumatic backfilling and
not likely to be used at active room and pillar faces.

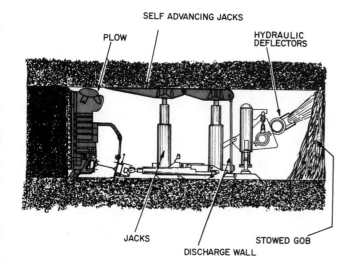

FIGURE 3.9 Cross-section through longwall face with lateral pneumatic stowing. (Singh and Courtney, 1974)

TABLE 3.1   European Coal Mines Using Pneumatic Lateral Discharge Units as of January 1, 1974 (Brieden)

Name of Mine	Year of Installation	No. Faces	Length of Face	No. Lateral Discharge Units
Nordstern (Germany)	1968	2	200 m each	95
Reden (Saar)		1	50 m	15
Erin (Germany)		1	45 m	10
Marienau (Lorraine)	1972	2	200 m	44
			300 m	66
Faulquemont (Lorraine)	1972	1	200 m	44
Folschiviller (Lorraine)	1972	1	200 m	44
9. Mai (Czechoslovakia)	1973	1	220 m	50
Anna (Poland)	1974	1	220 m	50
		10	1830 m	418

Hydraulic backfilling is most frequently used in steeply dipping seams. There is some evidence that it can be used in gently dipping conditions (as at the Miekowicze mine in Poland, see Figure 3.10), but water handling problems appear to prohibit its use in flat-lying seams such as those currently mined in the United States. The density of packing of the waste is superior to that obtained in pneumatic stowing, so mine roof disturbance and subsidence are more effectively minimized. The hydraulic method is preferred in Poland for use in the mining of thick coal seams (6 m or more in thickness) where the coal is extracted in three stages, of approximately 2 meters each. Although there are similarly thick seams and even thicker seams than these in the western United States, no such thick seams currently are being mined by underground methods. In Europe, these thick seams are mined in stages beginning with the uppermost seam or from the lowermost. In Poland the lowermost 2 meters is extracted and then backfilled hydraulically. Using the top of the previously backfilled material as the floor of the second mining the second 2 meters are then extracted in a similar fashion, at a distance (100 m or more) behind the bottom working face and backfilled in preparation for the third, final level. After the third extraction, the mine roof is allowed to cave. In this way, it has proved possible to limit to tolerable amounts the surface subsidence and associated damage above the mine.

Mechanical Backfilling

Mechanical backfilling is rarely used for underground backfilling in coal mines, although it has been used to a limited extent in Germany (discontinued in 1958), France (discontinued in 1962), Poland (in 1970 practiced in less than 1 percent of the mines), and in Belgium (in 1970 practiced in 2 percent of the mines) (5th International Strata Control Conference, 1972). In this backfilling method, coal waste is mechanically slung into a void. It has not proven effective in reducing subsidence because the packing density and the percentage of the void that is filled are considerably less than that of hydraulic or pneumatic systems. It could prove more promising as a method of waste disposal in certain applications. For example, in conjunction with a partial extraction system, coal waste could be moved from a borehole injection site to an area to be backfilled using mobile scoops and ram cars.

Hand Packing

Hand packing used to be a system popular in older coal mining methods in which "strip packs" were constructed manually by building a box-like structure of dry walls and filling them using rock available from local (controlled) roof falls. No waste was brought into the mine for the packs, so the system is of little interest as a possible method for disposing of mine waste. The packs served to support the roof over the caved area and localize roof falls along the caved area parallel to the face. Hard wood

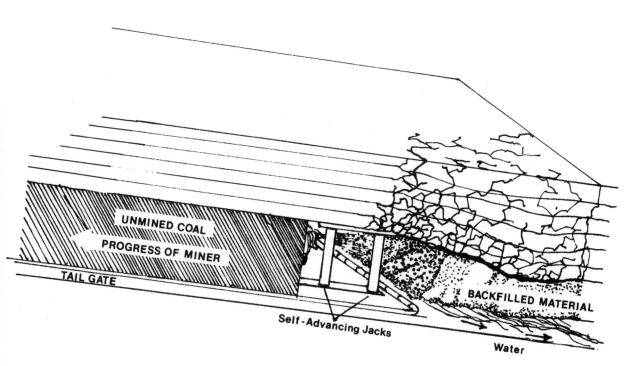

PLAN VIEW

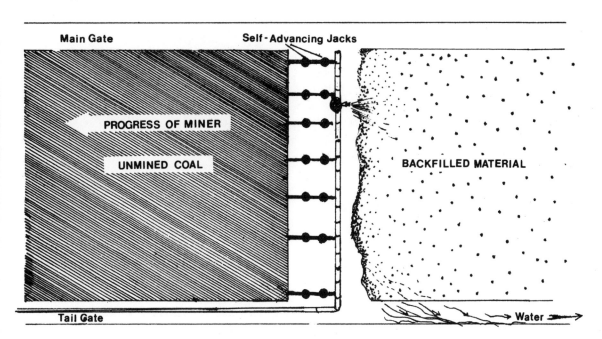

FIGURE 3.10 Hydraulic backfilling in Poland. The coal is extracted along a retreating longwall face. The seam dips gently,
allowing the water to drain from the backfilled material into a
sump where it is pumped through a pipeline to the surface. The
face is arranged to slope toward the tailgate and toward the
stowed area. The hydraulic backfill is discharged into the
mined out area behind the face.

or steel roof supports on the pack side of the face area served to limit roof collapse to the caved area, thus preventing it from advancing into the area at the working face. The system did not prove very effective for reduction of subsidence.

Strip packing is a relatively slow operation and although extensively used in the 1950's is no longer practiced to any extent. It gave way to complete caving, in which the roof is allowed to collapse fully behind mechanized roof supports. Caving eliminates delays in face advance that may occur with backfilling and thus is ideally suited for the high advance rates associated with mechanized longwall mining. However, caving does result in greater surface subsidence than any packing or stowing system.

Controlled Flushing (Hydraulic)

The term "flushing" is generally used to characterize systems for introducing backfill materials into voids in abandoned underground mines. In hydraulic flushing, a slurry of crushed coal mine waste and water is transported from the ground surface through boreholes and pipelines to the point of disposal in the mine (Figure 3.11). The water is then drained off and pumped back out of the mine leaving the waste material in the mine voids. The slurries generally contain 30 percent to 50 percent solids. The volume of water varies with pipe size, waste material size, and the desired pumping rate. Hydraulic flushing can be used to backfill flooded or dry mines.

Of the various types of hydraulic flushing, "controlled" flushing has been the most effective in backfilling abandoned mines. Controlled flushing employs men in the mine to place pipes and to control the distribution of the injected waste. It can be used in mined out portions of active mines, and in those abandoned mines where it is safe for personnel to install pipelines and perform the flushing operation.

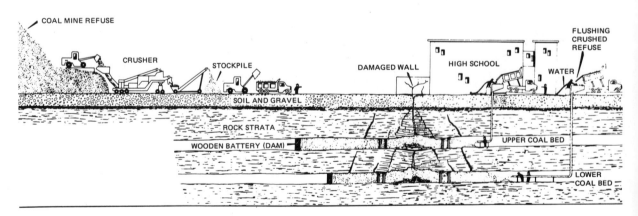

FIGURE 3.11   Controlled flushing (hydraulic).   (Bureau of Mines)

In controlled flushing, a pipe, generally 6 inches in diameter, is installed in a shaft or borehole from the surface to the mine level where it is connected by a large-radius pipe elbow to other pipes, either metal or plastic, which convey slurry to the areas to be backfilled. When a mine shaft is not open or safe, a large borehole (44-inch diameter or more) can be drilled into safe areas of an inactive mine to permit the entry of men and equipment. Wooden bulkheads and drain boxes installed downslope from the flushed area impound the waste slurry and permit the water to drain off (Figure 3.12).

Water pressure developed in the vertical section of pipe is sufficient to force the slurry through the pipelines in the mine. The distances to which the slurry can be distributed underground (under the pressure head in the vertical pipeline) varies with the depth to the workings, the frictional resistance of the pipe, the water-solid ratio of the slurry and the slope or dip of the mine. Generally 300-600 feet of horizontal distribution can be obtained for each 100 feet of vertical drop. Controlled backfilling should

FIGURE 3.12 Wooden bulkhead. This is an example of a pervious wooden bulkhead installed to contain slurry when controlled flushing is used to back-fill mine voids. It is also an example of the working environment in which controlled flushing takes place.

begin at the far limits of the area to be filled and progress to the point of final egress to the ground surface. Lower sections of a mine can be readily filled, but higher areas sometimes require the construction of bulk-heads to control the slurry.

Blind Flushing (Hydraulic)

Blind flushing employs no men underground and is used when it is impossible or dangerous to enter a mine because the mine workings are water-filled or have collapsed. In blind flushing, boreholes are drilled into the mine on a selected grid pattern, and the crushed waste is flushed from a hopper installed at the top of the borehole. The slurry flows into the mine with little control of backfill distribution. Placement of backfill continues until the void is filled and no additional slurry can enter the borehole.

An alternative method of blind flushing has been developed in which large volumes of slurry are transported into the mine at high velocity (Dowell, 1973). The process was designed to backfill flooded mines using sand or crushed mine refuse. The material is placed in suspension in a mixing plant (Figure 3.13) and pumped into the mined out areas through a closed pipeline system. The underground mine workings fill as solids settle out in response to changes in flow velocity. Peripheral to the area of turbulent flow at the base of the injection borehole, a doughnut-shaped mound of solids builds up toward the mine roof. The slurry velocity increases with the resulting restriction of available channel space and the suspended material is transported over the mound and deposited in quiet water on the outer slopes. Thus the mound builds outward (Figure 3.14). When one channel becomes blocked, another opens at the place of least resistance, and the process continues until a large area is filled. This technique has achieved a wide distribution of a large quantity of backfill from one borehole. At Scranton, Pennsylvania, in 1972 the maximum amount of refuse implaced through a single borehole was 204,000 tons, some of which carried more than 500 feet from the hole (Dowell, 1973). The average slurry was 17 percent by weight solids and the maximum amount injected was 61,000 cubic yards in two 8-hour shifts. A total of five holes were drilled, two of which were used to inject two-thirds of the material. Conventional blind flushing techniques would have involved 175-200 boreholes to achieve the same large areal distribution of backfill.

Hydraulic backfill requires large amounts of water. This may pose problems in some areas. Water for transporting fill into abandoned mines has been obtained from flooded sections of the mine or from dewatering activities in active mines. Water drained from the backfill or collected at a sump in the mine can be recycled into the backfilling system. Hydraulic back-filling of mines located above the water table may produce problems of increased stream pollution or slope stability. Hydraulic flushing into acid mine pools may exacerbate short-term water pollution problems (see section 2.3).

FIGURE 3.13 A mobile flushing unit. This blending
and pumping unit was used to backfill abandoned
mine voids in the Green Ridge section of Scranton,
Pennsylvania. The hopper and conveyor belt are
separate from this unit and must be moved from site
to site. The flushing unit can operate from the
back of the truck. (Appalachian Regional
Commission)

Pneumatic Flushing

     Experience with pneumatic flushing in the western Pennsylvania coal
fields has been limited to the pneumatic placement of fly ash.
     All but one of these demonstration projects has been blind injection.
In the only controlled flushing project access was possible to a mine beneath
a school addition. A cement block wall was built at the entrance to the mine
room underlying the school addition and fly ash was injected until the room
was filled (U.S. Department of the Interior, 1970).
     In pneumatic blind flushing, 6-inch diameter boreholes are drilled on a
grid spacing of 20-30 feet. The fly ash is transported in pneumatic tank
trucks which blow the fly ash into the mine through the boreholes at approximately 25 tons per hour. Photographs of mine backfilling have shown that

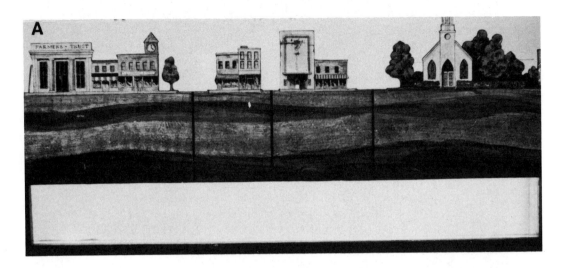

FIGURE 3.14a   A model showing various stages of deposition of fill material by closed system technique of
blind flushing using a single borehole.   A cross-section
of a flooded mine cavity underlying a developed area of
a city.

FIGURE 3.14b   Fill material suspended in water is piped
continuously from a blender to a pump at proportionately
high velocity and large volume into the mine void.   As
the mixture first enters the mine void from the injection hole, its velocity drops and the solid material
settles out to form a doughnut-shaped mound at the base
of the injection hole.

FIGURE 3.14c As the mound builds toward the roof of the mine cavity, the velocity of the suspended material increases across the mound and the solid fill material is transported over the pack.

FIGURE 3.14d The distance over which the fill will be transported depends upon the horsepower of the pumping unit, the geometry and slope of the mine cavities and the shape and continuity of barriers impeding the free flow of water.

FIGURE 3.15   A mine void backfilled with fly ash.
Note the roof contact and the dispersal of the
material.   (Pennsylvania State University)

the fly ash builds up in the mine void at a relatively flat slope (angle of
repose is approximately 10⁰) (Figure 3.15).   The spread of the material is
a function of the dimensions of the underground void.   In large rooms ash
spreads on the order of 30-40 feet from the borehole filling the voids with
reasonably good roof contact.   The properties of fly ash are quite different
from those of coal waste, but a pneumatic flushing system could be adapted
to handle coal mine waste.

3.2.2   Transportation of Backfill Material

The preceding discussion assumes that the backfill material is available
at the hopper entry to the stowing or flushing equipment.   However, delivery
of the waste from the coal treatment plant or mine dump may pose problems.
Transportation systems in coal mines are designed to move coal, men,
and supplies.   The systems use various combinations of equipment depending
on seam height, mining method used, and capacity required (see Draper, 1973
for a comprehensive description).   The coal transportation system is designed
for continuous movement of coal (and associated waste) from the face to the
surface treatment plant.   It may consist of locomotives hauling trains of
mine cars for miles to the preparation plant for rapid unloading into a
bunker using drop-bottom cars or rotary dump cars.   It may consist of a series
of conveyors, starting with a chain conveyor on the mining face itself, or

with shuttle cars, or scoops at the face, dumping the coal from a room into a surge feeder for delivery onto a secondary belt conveyor. Several such conveyors feed a main or "trunk" conveyor delivering coal to the surface.

Where a rail system is in use, trains returning from the surface sometimes are used to transport supplies. Conceivably, a system could be set up to transport mine waste for backfilling provided these cars could be delivered to the stowing machine hopper, unloaded, and returned to a coal filling point without interruption of coal production. Such scheduling would overload present transport systems and pose insurmountable problems in high production, mechanized mines in terms of synchronization of the backfilling and coal mining operations. A breakdown or delay in either would slow down the other and result in a loss in mine productivity.

If rail transport by rotary dump cars or drop-bottom cars is chosen to move the waste, dumping facilities will have to be excavated and installed underground near the stowing machine. In most U.S. mines this would involve the development of considerable additional head room at the dumping station and special feeder conveyors for transferring the waste to the stowing machine. With each move of the stowing machine new dumping stations and the conveyor systems would have to be relocated. In Europe the provision of waste dumping stations in tailgate haulageways is not a serious problem since head room in traditional European longwall mines differs from that in the United States because of at least two important mining practices:

1. Single entries rather than multiple entries are driven. Longwall advance rather than longwall retreat is still the more prevalent mining system. The entries are usually enlarged and maintained at a height of 8-10 feet to provide sufficient cross-sectional area for adequate ventilation at the face and to allow for access to the face even when convergence of the roof occurs with compaction of the caved and backfilled material.
2. Mine cars used for pneumatic stowing in European mines are generally smaller than mine cars now used in the United States.

Because United States longwall systems do not enlarge the entries as much as in Europe, and because longwall retreat is preferred to longwall advance allowing major convergence of the roof behind the retreating face when it is of less consequence than in advance mining, the head room in U.S. mines is invariably inadequate for rotary dumping. An entirely new system of rapid dumping and temporary storing of mine waste for stowing would have to be developed in U.S. mines. Special waste storage bunkers, with chain unloading of rail cars for some low head room situations have been developed in Europe and conceivably could be adapted for use in the United States mines.

Conveyor systems for coal transporation are popular in high production mines and have been suggested for transportation of refuse to pneumatic stowers. Although efficient for transporation, they are designed to be run in one direction only. Not only do the secondary conveyors overhang the trunk conveyors at transfer points, the pulleys supporting the belts are oriented so as to centralize the belt as it runs forward. This action is adverse when running in reverse. There are also other potential problems in the belt drive mechanism and in belt connections when driven in reverse.

In current operations the need for supplies in room and pillar mining (roof bolts, supports, etc.) and for mechanized longwall mining, can be met by relatively infrequent delivery during a working shift. Introduction of waste disposal would require the establishment of a continuous supply system--separate from coal transporation where this is done by conveyor.

Direct hydraulic or pneumatic transport of the mine waste in pipes from a surface bunker appears more feasible. As mentioned in Chapter 3, section 3.2.1 the Miekowicz mine in Poland uses such a system in its hydraulic stowing operation. Pneumatic conveyance of backfill through vertical pipes is used to transport wastes several thousand feet in metal mines. Although a potential problem, blockage of the system has not been a serious practical problem. Research conducted on the hydraulic and pneumatic transportation of coal to the preparation plant may resolve some of the potential problems of transportation into the mine for waste disposal. The use of booster pumps to maintain velocity of the waste in long horizontal stretches of mine roadway may be necessary.

The introduction of a waste disposal system into an operating mine will require a backfill transportation system suited for the particular conditions of that mine. In shallow mines, it might be practical to introduce waste into the mine through boreholes from the surface to a point near the area to be backfilled. Additional boreholes would be drilled as the disposal area became filled or as the mine face advanced. Such a system could encounter problems in populated areas or in hilly terrain, where in both cases, surface accessibility for boring could be limited and where surface transportation of the waste using trucks (or pipelines) on a continuous (24-hour) schedule would be undesirable.

3.2.3 Technical Benefits and Disadvantages of Backfilling

The principal benefits claimed for backfilling are reduced ground subsidence, disposal of waste, and reduction of ventilation leakage on longwall advance faces. The principal disadvantages claimed against backfilling are increased dust, fire and gas hazards, increased noise, more men underground, and roof control problems (discussed in more detail in section 3.2.4).

The volume of the surface subsidence trough resulting from "complete extraction" systems of mining is directly related to the real distribution and volume of the mine voids (see Appendix B). Backfilling in partially extracted areas by lending lateral support to pillars can prevent subsidence. Reduction in subsidence is of major concern when mining under developed areas or under rivers and bodies of water (either surface or groundwater aquifers). Backfilling in "complete extraction" and longwall systems will reduce subsidence, but cannot prevent it completely since the fill material is subsequently compacted by the weight of the overlying rock. Pneumatic stowing, the most widely used system in coal mines, reduces the maximum amount of vertical surface subsidence to about 50 percent of the extracted seam thickness without backfilling (British National Coal Board, 1966). Room and pillar mining in the United States has resulted in vertical subsidence of approximately 75 percent of the extracted seam thickness (Gray and Salver, 1971). Surface strains (the principal cause of damage to buildings and other installations) will be reduced correspondingly, but not necessarily enough to prevent damage.

Backfilling also reduces the disruption of strata in the vicinity of the mine workings, a feature that is of particular significance when mining multiple seams in close (vertical) proximity. It is often difficult to extract coal from a seam that has been disrupted by the effects of caving of underlying mined out seams. The method of multiple layer mining of thick seams in Poland is an extreme example of backfilling being used advantageously to maximize coal recovery from thick seams and multiple seams.

It is claimed in Europe that backfilling has improved ventilation of longwall advance faces (almost all U.S. longwall faces are retreat, not advance). Leakage or short-circuiting of air from the intake haulageway (usually the main entry or entries) to the return haulageway (the tailgate entry) via the voids in the caved area can be a serious problem in advancing longwall faces, especially as the haulageways become long (Figure 3.6). Backfilling may be a very effective way of reducing this leakage and insuring that an adequate supply of fresh air reaches the coal face.

There are several technical disadvantages of backfilling. Blockage of the pipe is an unresolved technical problem of pneumatic stowing systems. Care must be taken to avoid an excess of fines or clayey material from the waste as these tend to cause blockages in the stowing pipe system. Mechanized coal mining equipment produces a greater proportion of finer particles than older mining methods. Backfilling represents a major addition to the steps involved in the underground mining process by requiring a waste transportation and supply system to the backfilling equipment. If backfilling must be synchronized so as to keep pace with other coal face operations, the possibility of coal production loss due to pipe blockage, equipment breakdown, or lack of adequate waste supply could be considerable.

The high cost of backfilling was a principal reason for the decline of backfilling in Europe. Backfilling reached its peak in Europe during the late 1950's when the demand for coal was very high and it was necessary to mine as much of the available coal resource as possible while at the same time minimizing subsidence damage. This implied not only longwall mining of the coal seams, but also the extraction of contiguous seams. In order to restrict surface subsidence and facilitate multiple seam mining some form of stowing was deemed necessary. During the 1960's the demand for coal declined. Economic and social circumstances have dictated increased mechanization and a shift to a continuous 24-hour mining cycle. Less economical operations have been abandoned. Current mining practices include such economical measures as: (1) mining only the most profitable of the coal seams in a series (thereby reducing the total volume of voids contributing to subsidence but also reducing the resource recovery); (2) permitting caving instead of stowing, except in a few isolated situations; (3) introducing mechanized longwall systems in conjunction with the caving; (4) where feasible, designing surface installations to accommodate the resulting surface subsidence; (5) where not feasible, paying for subsidence damage rather than lowering production rates or institute a less profitable mining system involving backfilling.

A brief report supplied to this committee by the British National Coal Board represents the current status of backfilling in most European coal mining operations.

Stowing (backfilling) in Britain declined with the introduction of noncyclic mining methods, the principle of pneumatic longwall stowing depending very largely on mining in a cyclic manner. The cost of stowing which had always been very high in comparison with the other face costs, became excessively so as the labour cost proportion rose steadily in the post war years. The process of full face pneumatic stowing, when considered in its entirety, required considerable labour in spite of the low content of the actual face stowing operation. Power loading resulted in stowing material supply problems which gave further impetus to the decline as did the realization that waste caving had an almost universal application on our longwall faces, a fact which had taken considerable time to become established.

The consequence of these factors is that no full face stowing of any kind is currently practiced in British mines and only small quantities of nuisance dirt is being dealt with by means of pneumatic stowing, largely using the Schwartz Jet Stower of which about a dozen are currently in use. (British National Coal Board, January, 1973)

3.2.4 Health and Safety

Elimination of mine waste piles has obvious advantages of greater safety and improved environment on the surface. In assessing the overall merits of underground disposal, however, it is essential that its effects on the health and safety of underground miners be thoroughly examined.

Underground coal mining is a dangerous occupation, accounting for more than 100 deaths and 12,000 injuries annually in the United States, and exposing miners to the respiratory problems of pneumoconiosis ("black lung") and silicosis resulting from dust inhalation. In a system of underground disposal where a major fraction of the waste is deposited in the immediate vicinity of the working face in operating mines, the system could have a significant effect on the occupational hazards of the miner. In traditional European applications, backfilling was carried out after completion of the coal production cycle when the only personnel in the area were those engaged in stowing. With current high production systems, mining and stowing operations would be conducted simultaneously.

Factors that could be significantly affected include: (1) dust production, (2) gas control, (3) fire risk, (4) roof control, (5) noise, (6) personnel exposed to danger, and (7) hydraulic seepage and weakening of coal pillars and mine floors.

Dust Production

Pneumatic backfilling can produce a significant increase in the amount of respirable dust in the working area, especially at the start of backfilling

and if the backfilling is not sufficiently wetted. Backfilling is traditionally carried out so that face ventilation currents carried the dust away from
the operator. Where backfilling occurs simultaneously with production, miners
will inevitably be exposed to the airborne stowing dust. Systems to minimize this problem by wetting the stowing material as it is ejected are available but careful studies are required to determine whether the respirable dust
production can be brought within the stringent limits, of 2 milligrams per
cubic meter of air exposure to workers, set by the Federal Coal Mine Health
and Safety Act of 1969. This limit is difficult to achieve on longwall faces
even without pneumatic stowing.

Fire Risk

Some concern has been expressed that underground disposal of waste subject to spontaneous combustion (i.e., the counterpart of surface fires) could
lead to serious risk of noxious fume production or fires underground. This
possibility is again especially serious since any such gases would quickly
enter the underground ventilation system.

Spontaneous combustion results principally from the interaction of fresh
air with the combustible material so that, as with properly compacted surface
piles, there should be little danger of combustion once the stowed waste is
compacted by roof convergence. Hydraulic backfilling results in compacted
backfill without further compaction by roof convergence. The committee did
not find any published evidence concerning the question and suggests that it
be carefully examined.

Gas Control

In American mining practice, "bleeder entries" are required to be provided in connection with all full extraction mining systems, in order to prevent accumulation of methane gas in the caved area that develops behind the
retreating longwall (Figure 3.16).

Complete backfilling would restrict caving and substantially reduce the
system of voids created, although breaks would develop in the roof rock.
Gas under pressure above the seam level would thus still enter the breaks,
but would not be transported out of the extracted area, due to the high
resistance to airflow offered by the stowed material. The committee did not
resolve the question as to whether the ineffectiveness of bleeders in this
situation would result in any additional hazards along the working area.

European practice is not to provide bleeders. Although not required by
law, methane drainage is used in some instances. The gas is drained away
from ahead of the advancing longwall face through holes drilled into the
roof from the main haulageway and tailgate. The holes are connected under
suction pipes that transport the methane out of the mine. This system has
not been adapted for the retreating longwall method used in American longwall
practice.

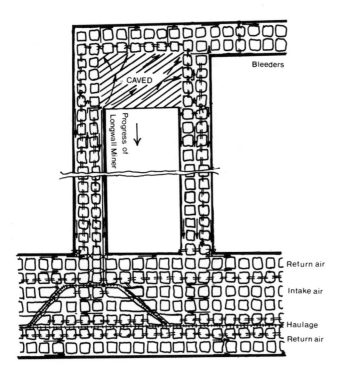

FIGURE 3.16 Ventilation of retreat longwall.

Roof Control

European experience suggests that complete backfilling tends to improve roof control in longwall mining, although the reverse may be true in seams susceptible to coal 'bumps' (not a large fraction) (Jacobi, 1966). Complete backfilling in room and pillar mining, if feasible, might adversely affect loads on pillars during extraction (see Appendix B). Moisture associated with hydraulic backfilling could result in some deterioration of roof and floor, and the general worsening of working conditions.

Noise

Experienced miners use the sounds of rock cracking and rock movements as valuable signals of impending danger. The noise associated with modern mechanized equipment has reduced dependence on the warning, but it is still valuable. Pneumatic stowing equipment is very noisy, especially near the discharge end of the pipe. The operator usually stands near this point in order to direct the backfill. The addition of other noise produced in the confined area of a coal face can only tend to increase potential damage to human hearing.

The current standards of noise exposure (90 dBA on an 8-hour basis) are criticized as too generous. Care must be taken not to introduce another component into the mining system which reduces compliance with the health and safety law.

Personnel Exposed

The need to operate stowing equipment in conjunction with other coal face operations inevitably would require that additional personnel be exposed to the hazards at the face. Additional personnel increases congestion and tends to raise the probability of accidents. It is generally agreed that, all other factors held equal (geological conditions, mining methods and management) the chance of an accident is related to the time of exposure to the potential risk. More workers in the face area increase the exposure, and, by virtue of congestion, the odds of risk.

Other Hazards Peculiar to Backfill Methods

Several of the stowing methods involve possible dangers peculiar to each. Thus, pneumatic stowing results in sparks produced when the waste rock is projected at high velocity from the stowing nozzle. The incendiary potential of the sparks was not determined but there appear to be no regulations opposed to stowing nor any record of accidents caused by sparking. This is an important topic that needs examination.

In hydraulic backfilling, there is danger that the barriers holding the stowed material in the backfilled area may burst and allow waste to flow onto the face. This problem becomes more serious in American continuous production operations, unless the backfilled areas are distant from the working face. The introduction of water into a mine may result in other, substantial problems. Under sufficient head, water will actually penetrate the coal, the roof and floor strata. The effective stress created by the water could reduce the strength of the pillars. Additionally, most strata associated with coal mines contain a high percentage of clay minerals, which may swell or be lubricated by the additional water. Slippery lubricated clays adversely affect the maneuverability of men and machines. The potential loss of strength in the roof strata and potential subsequent failure may be a significant safety hazard in some mines and should be critically analyzed and tested.

Overall, it appears that several unanswered questions of health and safety are raised when contemplating backfilling in conjunction with highly mechanized American coal mining operations. Careful study must be given to this problem. Modifications in current Coal Mine Health and Safety Regulations may be necessary to allow backfilling in operating mines.

3.3 Modifications of Present Mining Methods in Order to Backfill

Considerable experience has been gained with pneumatic and hydraulic backfilling behind longwall faces, and pneumatic and hydraulic backfilling in abandoned mines. In room and pillar operations extensive changes in present mining and transport methods would be necessary in order to allow the use of backfilling. Some of these had been practiced when mining methods were considerably less mechanized than today. Others are possible mining techniques where the technology is partially or fully developed. Still others reflect "drawing board" concepts based upon technology which is not currently available. These include techniques applicable to mining methods other than room and pillar such as in situ extraction. Some of these methods would circumvent the waste disposal problem altogether.

Creation of Partially Extracted Waste Disposal Areas

In a mining operation where only the wastes created as a result of current operations at a particular mine need to be disposed, it is probable that wastes will account for less than 25 percent of the volume of extracted material. Under this circumstance it may be technically and economically desirable to dispose of all the wastes in some particular section of the mine. An example of this would be an operation in which several mine panels would be developed by pillaring and the rooms left open as a repository for wastes produced from the coal mine. The wastes could be transported to the disposal area by rail or conveyor and pneumatically or mechanically packed into place. Such a system would have the advantage of not slowing coal production because of a lack of backfill material. A disadvantage of this scheme, as with all partial recovery mining methods, is that the pillars in the disposal area could not be recovered.

Secondary Recovery of Pillars after Flushing from the Surface

Secondary recovery of pillars after flushing was accomplished in the Anthracite region when mining methods were not mechanized and not regulated by current laws (U.S. Department of the Interior, 1912, 1913a and b). In 1956, the Bureau of Mines with the cooperation of an anthracite mine near Wilkes-Barre, Pennsylvania, also extracted pillars after flushing (U.S. Department of the Interior, 1956a, Williams, personal communication). The mines were reentered after flushing, and the pillars extracted by conventional cut, shoot, and loading. The resulting voids were also in turn backfilled by flushing (Figure 3.17). Some rehandling of the wastes in the original cross-cuts was inevitable. An important feature of the mine was the existence of a thick, competent, sandstone roof overlying the coal.

Aware of this success it is tempting (but generally misleading) to expect its possible application to bituminous room and pillar mines. Sections of the mine would be partially extracted, then barricaded and filled using hydraulic flushing. Shallow mines could be filled from the surface--in other cases the fill could be introduced at seam high points and water removed at low points. Once backfilled and solidified, the pillars within the section would be mined. The mine layout could be arranged to minimize rehandling of backfill, for example by a system of rectangular pillars with minimum width cross-cuts.

From a materials handling point of view, this backfilling scheme has the advantage of not interfering with routine daily mining operations and of using separate crews for backfilling. However, the system has several drawbacks. When first introduced, the waste material simply fills (and not

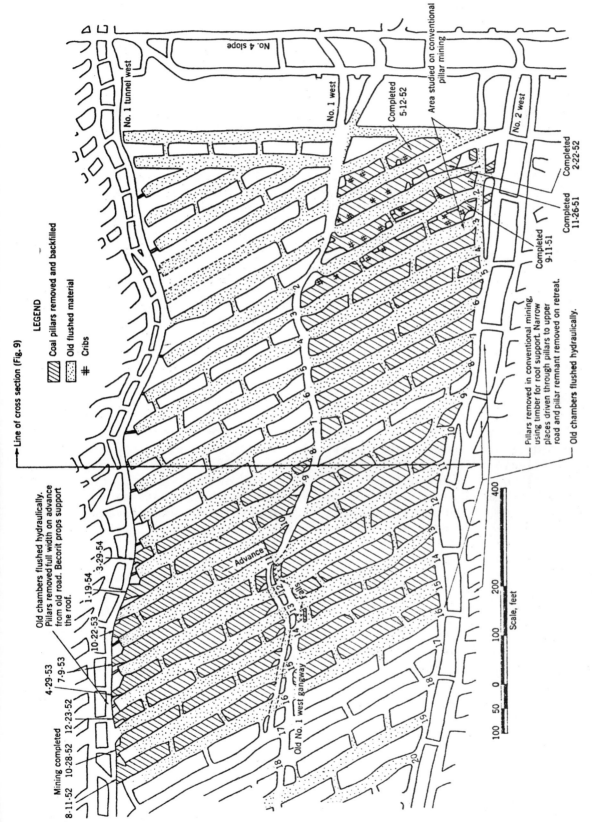

FIGURE 3.17 Secondary recovery of pillars after backfilling—the Anthracite Region. (U.S. Department of the Interior, 1956, p. 12)

completely) the rooms and carries none of the weight of the overlying rock--this being supported by the pillars. As noted in Appendix B, the ability of backfill to support an applied load develops progressively as it is compressed (Figure B.3, Appendix B). Extraction of the pillars will throw increasing proportions of the overburden weight onto the fill (Figure 3.18). The overburden load due to the weight of overlying rock remains constant (line AB). Initially the entire load (OA) is carried by the pillars. As pillars are removed the roof sags and converges onto the fill. After a sag equal to OF, for example, the total load will be redistributed such that amount BC is carried by the fill and amount CF by the pillars. Note that a sizable amount of the total convergence must occur before the fill carried a sizable portion of the load. In the St. Joseph Lead Company mines in Missouri, where the ore is sufficiently valuable to warrant substantial expenditure to permit removal, concrete replacement pillars have been used as "fill." They contain flat jacks that can be inflated to exert pressure against the roof to insure that the fill takes load at small deformation. Shifting of the load gives rise to shear stresses and substantial deflections in the roof layers. Unless these are competent and able to sustain the stresses, poor roof and hazardous working conditions are likely to develop. The amount of roof deflection could be as much as 20 percent or more of the pillar height for hydraulically placed fill and appreciably more for pneumatically placed fill.

Flushing of the secondary mining voids, where required, would require separate boreholes, significantly raising the overall costs, probably to a prohibitive level except perhaps in a very shallow mine under level terrain. If roof conditions are such that a significant fraction of the pillars must be left unmined, it would become even less economical.

Rehandling of a percentage of the material that had been initially flushed into the mine would be inevitable particularly with multiple entry mining systems and obligatory cross-cuts between them. As much as 40-50 percent of the fill material would be rehandled.

Because the method appears prohibitively expensive as well as dangerous under most current U.S. bituminous coal mining conditions, the committee does not consider secondary recovery a mining method that could alleviate the waste disposal problem.

Backfilling in Room and Pillar Operations with Complete Recovery as Part of the Mining Cycle

As already noted, room and pillar mining generally involves two stages: (1) mining of rooms; (2) extraction of pillars. In the second stage it is

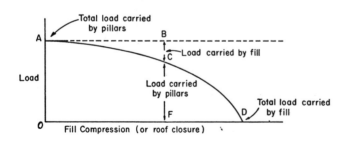

FIGURE 3.18 Distribution of overburden load between pillars and fill.

usual to allow the roof to cave behind the workings (Figure 3.3). A stepped pattern of extraction is developed to promote caving at some distance behind the pillar extraction line, thus providing some protection to the miner.

As in longwall retreat mining, the voids created during pillar extraction are "behind" the mining region, and it seems logical to suggest that backfilling into these voids could be accomplished without adverse effects. Indeed, if the roof is competent enough to allow a sizable part of the mined out area to remain open for a long period, it would be possible with existing technology to stow pneumatically from entries or rooms, other than those being mined, projecting the fill into the voids from a supported area. Projection of the waste to distances of the order of 50 feet into a void is considered possible in seams of moderate height (5 feet or more). However, a pneumatic system in a room and pillar operation would be more complicated than in the comparable longwall retreat system.

In most operating U.S. mines, the roof is relatively weak and caves very quickly after each segment of a pillar is extracted. Where this occurs it would be virtually impossible to use current, pneumatic backfilling systems. Modified undeveloped systems in which the stower is operated close to the continuous mining machine would increase congestion at the face, reduce maneuverability, and increase ventilation and safety hazards.

The system has several overwhelming disadvantages. Not only would it only be applicable under a typical U.S. mining condition, be awkward to maneuver and potentially increase safety hazards, it is not based on technology that is currently available or able to be modified. Even if practical, it would be a very expensive operation due to high equipment costs and extensive labor costs incurred by frequent moves.

Foam Encapsulation of Fill Material

The use of "foams" in underground mines was first investigated in 1954 by the U.S. Bureau of Mines in an attempt to apply various expandable plastics to sealing broken ground and limit access of air to fires in abandoned workings. From this first investigation many uses have been found for foams underground and considerable information has been developed that indicates more uses are possible (Mines Safety Appliances Research Corporation, 1973).

The term "foam" covers a wide variety of materials from detergent or soap foams, protein, rubber, and an extensive variety of resin base foams to the foams of cellular inorganics, cement, gypsum, anhydrite, and silicate. Of these, detergent and polyurethane foams have been used in underground mines. Cellular concrete is being considered as a material for building walls in mines (packs or stoppings) but no underground demonstration work has been conducted. It is normally more fluid than that used in concrete construction and can be pumped over horizontal distances of 1,000 feet. Anhydrite (with setting accelerators) is used as a mixture with mine waste for construction of high strength mine packs in underground coal mines in Germany.

The utilization of foams in returning coal mine wastes underground offers several advantages. In pneumatic stowing, foams could be used to suppress dust and to encapsulate the stowed material thereby preventing spontaneous combustion and acid mine drainage. Foam or cellular concrete with coal waste as the aggregate would be stronger and less compressible than untreated waste and might provide enough overburden support to enable safe secondary recovery of coal pillars in relatively shallow mines.

A disadvantage of foam encapsulation of waste is that it represents a substantial added cost to the disposal process. In the case of cellular concrete it probably would not be economic to consider making a mix strong enough to support the overburden. Even using foam concretes there still would not be sufficient waste material to backfill an entire mine since waste is only 25 percent of the mined volume.

Hydraulic Mining Systems

A recent research announcement of the U.S. Bureau of Mines (July, 1973) lists one new contract on the feasibility of hydraulic transportation of coal and three new contracts on the hydraulic cutting of coal. It is reasonable to assume that there will be in the near future prototype systems for underground hydraulic mining and handling of coal. If this assumption becomes reality, there will be underground mines designed to handle volumes of water as part of the coal production cycle. Hydraulic transport does not imply loose water in the mining area nor would hydraulic cutting of coal necessarily result in large quantities of additional water at the mine face because the high pressure jets utilize comparatively little water. Hydraulic backfilling, on the other hand, requires relatively large quantities of water. Although such water could be recycled into the system, its drainage from the backfill may not be continuous enough to guarantee the flow necessary for a continuous hydraulic system. Moreover, the problems of controlling water in nearly horizontal seams would be very difficult to overcome. Therefore, it cannot be assumed that hydraulic backfill can be added to hydraulic mining systems without significant redesign of the system. However, with the development of systems that handle water underground hydraulic backfilling would be more feasible. It is impossible to postulate equipment and manpower needs for such a backfill system until the mining part of the hydraulic system is developed.

Underground Coal Washing

Inasmuch as coal mining and backfilling are primarily materials handling problems, any change that decreases the amount of rehandling should be considered. One suggested change is to locate preparation plants in the mine. Waste could then be replaced in the mine voids by backfilling techniques without long transportation hauls. It is unlikely that this could be accomplished safely because of hazards from methane. Also, it is impractical because of the impermanence of coal mine openings. A variation of this idea would be to design a continuous mining machine that contains the primary cleaning circuit within it. If it were possible to remove some percentage of the waste material during retreat mining, then it could be backfilled immediately. The size and weight of present cleaning units make this plan strictly conjecture and currently impractical.

Automated Mining Systems

Significant research has been conducted in automating mining machines in order to reduce the exposure of miners to hazardous face conditions. The most successful projects, for the short term, do not remove operators from the mine. They are removed, however, from the hazardous face area (usually accepted as being with 25 feet from the face of coal being cut or supported) and placed behind an instrument that remotely controls the mining machine. The operator's own senses are used to perceive conditions and their commands are relayed to the machine by wire or radio control.

More complicated schemes of automation have been proposed. These involve remote sensing of mining conditions and long-distance remote control of machines. Although the operators would be removed from the coal face by these schemes, the increase in complexity of machinery would result in greater exposure of maintenance personnel to the more dangerous face situations. As problems of sensor reliability and durability are overcome, it is conceivable that automated mining could be coupled with nonexplosive atmospheres to reduce personnel accidents.

In principle, backfilling equipment could be incorporated into an automated mining system, but might tend to make the already complicated machinery even more cumbersome.

In Situ Extraction by Gasification and Solvent Extraction

Techniques for in situ gasification of coal have been receiving increased attention during recent years. In situ gasification involves converting coal into a combustible gas by partially burning the coal while passing a stream of air and steam through the coal seam. The steam reacts with the coal and creates a gas that then is forced to a series of collection wells. In situ gasification would have the advantages and disadvantages of normal gasification. It would produce relatively "clean" fuel and theoretically make low-grade coals and lignites economical deposits. In situ extraction would have the further advantage of doing away with the mining process itself and hence with the problem of disposal of mine wastes. The process may have severe disadvantages such as subsidence, pollution of groundwater, and aquifer interruption where the coal bed itself or overlying strata are major water aquifers. It is difficult to postulate whether these disadvantages would be more or less pronounced than in conventional mining.

Inasmuch as some of the proposed coal desulfurization methods involve the use of organic solvents to dissolve selectively the hydrocarbon fuel, there is reason to investigate the use of these solvents for in situ extraction of coal. The major problems of such a scheme are solvent cost and recovery, and, as with any solution-mining scheme (salt, or sulfur for example), control of solvent leakage to the groundwater and subsidence of the resulting cavity. A major advantage is the elimination of the handling of coal mine wastes and pollutants. Such a scheme would eliminate the problems of both surface and underground disposal.

CHAPTER 4
ECONOMIC FEASIBILITY OF UNDERGROUND DISPOSAL OF WASTES

SUMMARY

Estimates of the economic costs of backfilling are made based upon (1) actual costs of European experience in active mines, (2) actual costs of backfilling abandoned underground mines in the United States for subsidence control, and (3) engineering estimates of hypothetical situations using typical costs of U.S. mining operations.

The costs of environmentally acceptable surface disposal and underground disposal vary critically with each mine site, but the cost of underground disposal will almost always be higher than that of surface disposal.

Underground disposal systems may not eliminate surface disposal of certain fractions of the coal waste. Pneumatic systems are not designed to transport and dispose of the finest fraction of the coal waste; this would still be disposed on the surface. Hydraulic disposal in abandoned sections of a mine cannot take place until whole sections of the mine have been developed and mined. In the interim the waste must be disposed on the surface and the cost of such interim surface disposal systems assumed by the company.

The capital outlay for the stowing system and maintenance of it accounts for 1/2-3/4 the cost of the total underground disposal system. Other costs include the crushing, transportation, and short-term surface storage of the waste, the long-term surface disposal of the mines, and the cost of the injection boreholes. Factors that greatly affect the cost of backfilling include the thickness of the coal seam (i.e., the volume of the void), the capability of the system to totally backfill an area, the prevailing surface land use pattern, the cost of temporary disposal sites, the associated costs of transportation and the costs of neighborhood inconvenience, and the indirect costs due to loss in production in the mine. Under certain conditions the costs due to losses in production caused by disruption of the mining cycle could double the cost per ton of underground disposal.

Introduction of concurrent backfill and extraction systems into an operating mine could seriously reduce production; but where the backfill and extraction are separate, there should be no direct productivity loss.

If backfilling were considered necessary, new mining methods undoubtedly would be evolved for new mines to allow greater efficiency than could be achieved by the addition of backfilling to today's ongoing operations.

4.1 ECONOMIC COSTS OF BACKFILLING

There are no reliable cost estimates for underground disposal of coal mine wastes in the United States. The committee relied upon three sources for its cost estimates: European experience in underground backfilling, backfilling for subsidence control in abandoned mines in the United States, and estimates based on typical costs of current United States mines with projected costs for hypothetical modifications to include backfilling. These estimates were compared with those of manufacturers of backfilling equipment and with estimates by academic and research institutions.

European costs are based on actual backfilling experience in operating mines in Britain, Germany, Poland, France, and the Netherlands (see section 4.1). These mines have distinctively different geologic settings and mining conditions than those of the United States. European mining conditions tend to increase the cost of producing coal and decrease mine productivity compared to that in the United States. The cost of land and energy are higher in Europe than in the United States and the costs of surface damage from subsidence are also higher. The cost of labor is lower. Although the opportunity cost of capital for the private sector is lower than that in the United States, the European Coal and Steel Community of the Common Market provides low interest loans on an "advantage basis" for modernization of mines. Typical interest rates are 5-5 1/2 percent for the first 5 years and 7 3/4-8 1/4 percent thereafter. Because of these widely disparate economic, social, and physical conditions, European costs for backfilling should be compared to the total cost of European coal production before being compared on a per ton basis to U.S. costs.

Data used in this report from the United States are from actual operations for subsidence control in areas above abandoned mines (see section 4.2). These data reflect an engineering objective of subsidence control, rather than an engineering objective of disposal of wastes underground. The subsidence projects were designed for only a single period of use and thus lack the integrated operations of an active mining process, reducing waste output and costs, or of utilizing otherwise under-utilized equipment, men, and other factors of production. Many factors have influenced the cost of subsidence control projects. In addition to technical matters, the mere involvement of government sponsorship increases costs to levels higher than if the work were performed for private industry. All federally supported work must be done using the Bacon-Davis Act wage scales which are usually higher than the prevailing local mining wages. In addition, state, federal, and consulting engineering, overhead, and government procurement procedures increase costs. Most subsidence control projects are in an urban environment where costs for utilization of heavy drilling, crushing, trucking, and pumping equipment are increased. In most subsidence control projects the depth of drilling is substantially less than that of typical active bituminous coal mines. Only one-third of this footage of drill holes is cased. These factors reduce costs per borehole but are offset by observation holes drilled in abandoned mine projects to ascertain the success of the backfilling. Most of the abandoned mines are flooded; consequently, there are no costs for obtaining water, for pumping away water, or for treating water used for flushing (see section 4.2).

TABLE 4.1 Committee Estimate of Backfilling Costs

	Range Cost/ton of Coal Mined	Cost/ton of Coal Mined	Cost/ton of Waste
European underground disposal costs	$.50-$5.00	$1.00-$3.00	$3.00-$9.00
U.S. abandoned mines underground backfilling costs	--	--	$3.00-$9.00 average $4.00
U.S. "environmentally acceptable" surface disposal costs	$ .5-$.50	$ .25	$1.00
Minimum hypothetical underground disposal costs with pneumatic backfilling	$1.00-$5.00	$1.00-$3.00 minimum	$3.50-$5.00
Minimum hypothetical underground disposal costs with hydraulic backfilling	$1.00-$5.00	$1.50 minimum	$5.00

NOTE: This chart assumes that 25 percent of the raw coal is waste and must be disposed of either on the surface or underground. It also assumes that the physical conditions necessary for disposal are available near the preparation plant. Some operators that sell raw coal have virtually no costs associated with refuse disposal. In other areas conditions may be so adverse that "environmentally acceptable" disposal may be economically prohibitive. The cost estimates for underground disposal are those for the most advantageous conditions and are minimum costs. The cost estimates for surface disposal are for average conditions using the best available techniques and machinery.

Cost estimates from manufacturers of backfilling equipment are lower than those of contracts for subsidence control or by the committee. Such estimates assume that the backfilling operations will work as efficiently or more efficiently than any other part of the mining operation.

The committee estimates of both surface disposal costs and underground disposal costs are essentially a compilation of the incremental costs associated with each step of the disposal operation (see sections 4.3, 4.4 and 4.5). Such a composite of operational costs considers disposal of waste an additive operation rather than an integral step of mining operations. This is probably true for active mines that change over from surface disposal to underground disposal or for mines planned in the near future. With time, more efficient machinery and mine design and waste disposal integrated systems may tend to lower costs.

Table 4.1 is a summary of the cost data examined by the committee in the following sections.

4.2 EUROPEAN COSTS

Coal mining in Great Britain, Belgium, the Netherlands, France, and West Germany has been declining in recent years (see Table 4.2). Backfilling of waste materials in underground coal mines has also declined in Europe over the years not only due to the decrease in coal production, but because the costs of backfilling have exceeded the economic benefits of subsidence control. Stowing by pneumatic, hydraulic, and mechanical means has markedly decreased in all the countries referred to above. Higher costs for labor, materials, and equipment will tend to further decrease backfilling. Higher values for land and higher values placed upon environmental quality, and hence, higher costs for surface disposal tend to counter this trend.

European costs for backfilling vary considerably on a per ton basis and as a percentage of the total production cost of coal (see Table 4.2). The costs presented in Table 4.2 do not include the loss of mine productivity caused by backfilling operations. These costs are assumed to be significant in active mines in the United States (see sections 4.4 and 4.5). Because productivity is lower in Europe than in the United States, the effect of hours lost in productivity due to backfilling will be proportionately less costly in Europe than in the United States. However, on the assumption that a back-filling machine would inhibit the rapid progress of a longwall face, the Aberfan Tribunal concluded in 1967 that backfilling for waste disposal was not economically feasible in Great Britain (see Appendix D). Recent advancements in backfilling technology might influence this decision (Brieden, and Radmark, personal communication).

Poland is an exception to the general trend in Europe of rapidly declining use of backfilling. Poland has stringent surface disposal criteria and requires in some areas that underground coal mine wastes be returned underground. The reason for the prevalence of backfilling in Poland was not ascertained nor was the cost of Polish mining and backfilling operations.

Table 4.2 reflects considerable uncertainty concerning the costs of waste disposal in Europe. Even the estimates of European costs for coal production are uncertain. Cost estimates for actual surface disposal practices and associated costs for land acquisition, for complying with regulations for land use management and aesthetic considerations, compensation for subsidence damage, and the cost of water pollution control have not been compiled in a systematic way to permit international comparisons. For this reason it is not possible at this time to compare European production costs, social costs, or disposal costs with those of the United States.

4.3 UNITED STATES EXPERIENCE IN BACKFILLING ABANDONED MINES FOR SUBSIDENCE CONTROL

Coal Refuse as a Stabilizing Medium in Abandoned Mines

A number of mine stabilization projects have been completed in the Northern Anthracite Field of Pennsylvania using anthracite breaker refuse crushed to minus one-half inch as the fill material (Figure 4.1a, b). The cost of the projects were borne jointly by federal and state governments and were part of a program called Operation Backfill. Controlled flushing and blind flushing techniques of backfilling were used to fill mine voids.

Conditions in abandoned mines are very different from those found in active mines. Some of these differences make certain phases of underground backfilling operations cheaper in abandoned mines than in active mines, e.g., when blind flushing is done from the surface, no miners are needed underground. Other factors make backfilling of abandoned mines more difficult, less successful, and more expensive than active mines, e.g., lack of access, roof falls blocking the flow of material, and limited materials handling facilities on the surface when working in urban areas.

Technical considerations influencing costs of backfilling abandoned mines include the size and geometry of the area to be filled; remote observation systems; the height, interconnection and volume of voids; the cost of backfill material and transport medium, the cost of transporting the material to the injection site, the size to which the material is crushed, and the method of injection. Nontechnical consideration factors that tend to bias costs include short-term contracts and increased labor costs.

Table A of Appendix C presents cost data and volumes of coal waste utilized in 27 federal and Pennsylvania backfilling projects from 1956 through 1972. The costs listed have not been updated to 1974 costs. The projects range in cost from $1 to $15 per ton of waste. From these figures, it is not possible to draw conclusions on the economic advantages of different methods of flushing (i.e., blind or controlled). Controlled flushing necessitates the drilling of a 44-inch diameter hole which costs approximately $450/lineal foot. In blind flushing only a 12-inch diameter hole is drilled but it is necessary to monitor the success of the projects by several remote observation holes that each average $40/lineal foot. The backfill material and the water used in such systems are provided without charge for federally funded projects. A breakdown of the cost of one hydraulic, blind flushing, federally funded project is provided in Table 4.3.

TABLE 4.2 Trends in European Underground Coal Production and Production Costs

	Great Britain		Belgium		The Netherlands
Production by year (in millions of metric tons)	1951: 214 1958: 202 1962: 191 1966: 168 1968: 184 1970: 135 159 1972: 132		1951: 30 1958: 27 1962: 21 1966: 18 1968: 16 1970: 13 1972: 12		1965: 12 1968: 8 1970: 5 1972: 4
Average waste (%)	10-20%		40%		missing
No. active mines	1960: 698 1965: 504 1970: 299 1972: 289		1960: 75 1965: 54 1970: 24 1972: 20		1960: 12 1965: 11 1970: 5 1973: 4 phase out by 1975
% of mines that backfill	Complete* Partial** 1954: 5% 76% 1962: 3% 54% 1966: 2% 37% 1970: 4% 1973: *by hand, mechanical, pneumatic, and hydraulic methods **road packing		Pneumatic Hydraulic 1951: 2.5% 5% 1958: 2.5% 15% 1962: 5% 19% 1966: 6% 13% 1970: 4% 12%		1973: 25% (1 mine)
Total production cost	1970: £ 5.80 = 1972: £ 7.66 =		1970 = $35		1972: $40-50
Cost of backfilling per ton of coal	pneumatic $1-$2.50		pneumatic $1-$3.50		pneumatic $2
Costs of surface damage from sub-sidence where assessed	without backfilling: $2 with backfilling: $.05		without backfilling: $2 with backfilling: $.20		with backfilling: $.50-$1

TABLE 4.2 (Continued)

	France		Germany		Poland	
	Pneumatic	Hydraulic	Pneumatic	Hydraulic	Pneumatic	Hydraulic
Production by year (in millions of metric tons)	1951: 55 1958: 59 1962: 54 1966: 52 1968: 50 1970: 40 1972: 36		1951: 135 1958: 149 1962: 141 1966: 126 1969: 117 1970: 111 1972: 108		1962: 109 1964: 116 1970: 139 1971: 144	
Average waste (%)	30-40%		25-35%		15%	
No. active mines	1960: 95 1965: 70 1970: 49 1972: 40		1960: 146 1965: 107 1970: 69 1972: 61		1973: 78	
% of mines that backfill	1951: 2% 1958: 8% 1962: 10% 1966: 10% 1970: 8% 1972: total	14% 18% 16% 19% 20% 28%	1951: 21% 1958: 30% 1962: 24% 1966: 16% 1970: 11% 1972: 10%	-- .5% 1% 2% 1% 1%	1962: 39% 1964: 42% 1970: 41% 1972: 34%	12% 3% 3% 3%
Total production cost	1970: $30		1970: $30-40		1970: $14	
Cost of backfilling per ton of coal	Pneumatic $3.50-$5.50		Pneumatic: lateral discharge: $.75-$1.50 standard pneumatic: $1.50-$3.50			
Costs of surface damage from subsidence where assessed			without backfilling: $2-$2.50 with backfilling: $.15-$.65			

SOURCES: European Community for Coal and Steel (Production costs); 5th International Strata Control Conference, 1972 (Europe); Gluckauf (Germany, Britain, Belgium, the Netherlands, France, and Poland); Industrie Minerale Mine (France and Britain); Singh and Courtney, 1974 (Backfilling costs and production costs); Steele, 1973 (Poland); National Coal Board (Britain).

FIGURE 4.1a Scranton, Pennsylvania, November 20, 1966. The Von Storch (left) and Eureka (right) mine waste piles. These piles were the source of backfill material for the Appalachian Subsidence Projects 1 and 2. The Eureka Bank was the source of backfilling material for the Public Health and Safety Project No. 2, Tripp Slope. (U.S. Bureau of Mines)

FIGURE 4.1b Scranton, Pennsylvania, Summer, 1973. After several subsidence control projects, much of the Eureka bank and part of the Von Storch bank had been used for backfill. (U.S. Bureau of Mines)

Backfilling of Coal Mines Using Materials Other than Coal Mine Wastes

Fly ash has been placed underground for disposal and considered for mine roof support, for pillar support, and for extinguishing fires. Sand and gravel have also been used for backfilling mined areas. The properties of these materials particularly fly ash differ considerably from coal mine wastes. Fly ash particle size (usually less than 1/200 inch) and shape (usually spherical) determine that this material will have properties vastly different from coal refuse. Fly ash has an average angle of repose of ten

TABLE 4.3

A. Projected cost breakdown of a federally contracted hydraulic backfill project of an abandoned mine in an urban area.

Crushing, stacking and delivery to hydraulic system

300,000 tons at $2.65/ton contract charge

3 shifts/day   7 days/week   50 weeks

Total cost $796,000

Surface transportation from the waste pipe to injection holes by slurry pipeline, 5 injection holes,

300,000 tons at 250 tons/hour and 500,000 gallons of water/hour

1 shift/day   5 days/week

Total cost $900,000 or $3.00/ton

Exploratory drilling and monitoring (20 holes) $49,000
TOTAL $1,745,000 or $5.82 per ton of waste

B. Cost estimates for pneumatically backfilling undermined urban areas with fly ash for subsidence control:

Interval between Ground Surface and Abandoned Mine	Cost/Acre	Cost/Cubic Yard
30-60 ft	$60,000	$15.00
60-100 ft	$70,000	$17.50
100-150 ft	$90,000	$22.50

C. Cost of using fly ash to extinguish mine fires:

	Cost/ton ash
Dry fly ash injection (12 projects)	$2.56
Water-fly ash slurry (3 projects)	$7.83
Sand and fly ash slurry (3 projects)	$3.94

TABLE 4.4 Committee Estimate of Costs for Surface Disposal

Operation	Total Amount or Number	Fixed Costs Total/ yr.	Cost/Ton Clean Coal	Costs Related To Coal Production Total/ yr.	Cost/Ton Clean Coal	Costs Related to Tons Refuse Handled Total/ yr.	Clean Coal	Totals Total/ yr.	Cost/Ton Clean Coal
Land Acquisition	$250,000								
Total		$ 7,500	$ .004					$ 7,500	$ .004
Disposal System									
Capital	$708,900	$183,000	$ .088						
Labor	4 man shifts/days								
Oper., supply & power				$ 60,700	$ .029	$207,000	$ .100		
Total								$450,700	$ .217
Reclamation--									
Contract charge									
Topsoiling						$ 33,500	$ .016		
Revegetation						$ 1,500	$ .001		
Total								$ 35,000	$ .017
Disposal of Fines									
Capital	$500,000	$129,000	$ .062						
Labor	50 man-days/year								
Oper., supply & power				$ 3,300	$ .002	$ 3,500	$ .002		
Total								$135,800	$ .066
Subtotal		$319,500	$ .154	$ 64,000	$ .031	$245,500	$ .119	$629,000	$ .304
Productivity loss (no loss)									
TOTAL									$ .304

NOTE: A further breakdown of costs is presented in Appendix C, Table D(2)--2,070,000 tons of clean coal per year; 690,000 tons of refuse produced per year; 621,000 tons disposed on a waste pile; 69,000 tons disposed in a slurry pond.

degrees versus over thirty degrees for coal refuse. Fly ash, when mixed with lime, sets up when moistened--unlike coal refuse. While coal refuse is usually a generator of acid leachate, fly ash is most often alkaline. The flour-like quality of fly ash coupled with its spherical particle shape give fly ash unusual flow characteristics. Typically, large voids may be filled with fly ash through a small diameter boring.

Cost estimates for backfilling with these materials cannot be compared directly with those of coal waste disposal. However, these materials are being placed underground by equipment similar to that which might be used to dispose of coal mine waste. Table 4.3 summarizes (a) federal projects that have used materials other than coal waste for surface subsidence control, (b) cost estimates for pneumatic backfilling of fly ash in an urban area, and (c) costs of using fly ash to extinguish mine fires.

4.4 ESTIMATED SURFACE DISPOSAL COSTS

Current costs of coal waste disposal in the United States vary within a narrow range and depend primarily on the choice of materials handling systems, topography of the mining operation and the distance between the source of waste and the disposal site. Almost exclusively, waste disposal for underground mining operations is currently accomplished by surface disposal on waste piles near the preparation plant.

Current practices of surface disposal consist of dual waste disposal systems, one for the coarse waste and one for the fine waste. Coarse waste is usually handled by truck, belt conveyor or aerial tram or a combination thereof and deposited directly on the waste pile. Fine waste, approximately 10 percent of the total waste, is normally transported as a slurry by pipeline to an impoundment. If the fine waste is completely drained, it can be removed from the impoundment and deposited on the waste pile. More often it is left in the dry pond.

A total waste system usually consists of several component units; a conveyor belt for traversing steep grades, long distances and for large volumes, a truck for distribution to different areas, and a dozer and roller for grading and compacting. In some cases a large scraper (earthmover) will distribute, grade, and compact the waste in one operation. The establishment of vegetative cover as well as the grading and compaction of the waste is a part of environmentally acceptable waste disposal. Typical costs for currently operating total waste systems (belt, truck, dozer) range from $.15 to $.40 per ton of waste handled. Engineering estimates for installing a new system based on current costs of new equipment are considerably greater (see Table 4.4 and Appendix C, Table D-2).

By Truck

Hauling coarse waste by truck is perhaps the most reliable and efficient method used today especially in regions where the haul road grades are not excessive. Trucking waste for distances usually encountered in typical deep mine waste disposal systems (0.5 to 1.5 miles) is frequently done with large off-highway trucks which pay no license or fuel taxes.

By Belt Conveyor

Coarse waste disposal by belt is usually practiced in regions where terrain is too steep for trucks, but no steeper than 20°. This operation is usually accompanied by truck, scraper-loader and/or dozer to distribute, grade, and compact the waste at the disposal site. Costs of transporting waste by conveyor belt are quite sensitive to distance and amount of material handled. The capital cost for a belt conveyor is high, on the order of $200 per lineal foot.

Hydraulically

Hydraulic transport of waste is generally confined to the fine fractions. The coarse fractions (3-4 inches) have been transported hydraulically, but excessive wear of pumps and pipelines resulted in its discontinuance. The cost per ton-mile for pumping *only* is roughly comparable with trucking, but the cost of properly constructing approved impoundment dams for settling basins can make this part of the disposal costs very high. The cost of a dam varies with its volume (length and cross-sectional area). It is not possible to generalize the costs of various sites and operations.

Aerial Tram

Where the topography of the mining operation is too steep to permit the use of trucks or belt conveyors (greater than 20°), aerial trains or trams have been used. Most installations of this type are located in the steep Appalachian regions and, like belt conveyor systems, are usually supplemented by trucks, dozers and/or loaders to complete the distribution of waste material at the disposal site.

Other

Lorry cars, side dump cars, and car hoists have gained only limited applications in modern coal mining waste disposal systems. Cost data are not reported because of the unique and almost custom designed nature of these operations.

Among other costs which vary between great extremes are the cost of land and reclamation of the disposal pile. These costs are dependent upon location, topography, and local and state laws.

Typical Waste System

A typical surface waste disposal system could consist of: (1) A short belt (1500 feet) which takes the waste from the preparation plant to the surge bin. (2) A truck which takes the refuse from the bin to the disposal site. (3) A dozer with a roller for spreading, compacting, and reclamation (soil covering and contouring to specified slopes). The seeding of the pile and the planting of trees are usually contracted out.

An estimate of the cost of a surface disposal system designed to meet demanding environmental standards has been estimated by the committee in Table 4.4 and illustrated in Figure 4.2. A breakdown and explanation of the costs is contained in Appendix C, Table D.

4.5 COMMITTEE ESTIMATE OF PNEUMATIC BACKFILLING

In designing a hypothetical pneumatic backfilling system the committee tried to envisage a mine with representative conditions that would result in the most economically feasible backfilling system. The pneumatic backfill system designed is shown in Figure 4.3. Similar backfilling systems are used in Europe although the mine conditions are not nearly as favorable as those projected in Figure 4.3. Pneumatic systems can transport material horizontally approximately 2,000 feet without resuspension of the heavier material. Each corner that the pipe turns is equivalent to the frictional loss of energy in 200 horizontal feet of transport. In the committee's model, the horizontal capacity of the system is considered to be 2,000 feet including two right-angle turns, i.e., slightly greater than the maximum capacity of available technology.

The most economical pneumatic backfill system will dispose of the entire mine's refuse at a single working face or panel thus minimizing capital expenditures and labor costs. The committee projected backfilling behind a longwall panel, because pneumatic systems have proven feasible on European longwall panels, and because of unresolved problems associated with room and pillar pneumatic systems. The specific assumptions used in this projection are summarized below. Costs are presented in Table 4.5a, b, and c and further explained in Appendix C, Tables E(a, b, c).

Assumption: One ton of refuse fills approximately the volume of void created by the removal of one ton of raw coal. Waste production is assumed to be 25 percent of raw coal production (see Chapter 1). If a single longwall panel is to accommodate all the refuse from a mine, the longwall panel being backfilled will account for only 33 percent of the mine's production. In most mines there are several panels, and room and pillar extraction is used to mine other sections and to drive the entries of the longwall panels. Three examples of proportionate quantities of refuse to be disposed are given in Tables 4.5a, b, c. In Table 4.5a the volume of coal extracted from the long-wall panel is in equilibrium to that of waste produced by the entire mine. In Table 4.5b the amount of waste to be backfilled is only that produced by the longwall panel and is of considerably less volume than the volume of coal removed.

• It is assumed that the longwall panel averages 1,000 tons of raw coal per shift. Although the committee's model assumes that the longwall operates reliably at 1,000 tons per day, experience has shown that longwall panels operate sporadically--producing 2,000-5,000 T/shift when operating and virtually none for entire shifts on days when there is trouble with the machine or it is being moved to a new panel (see Appendix C, Figure G).

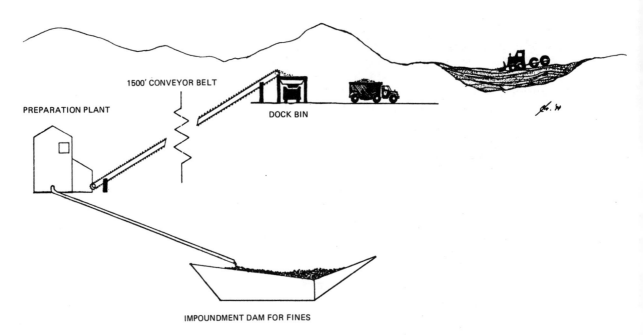

FIGURE 4.2   Surface Disposal System.

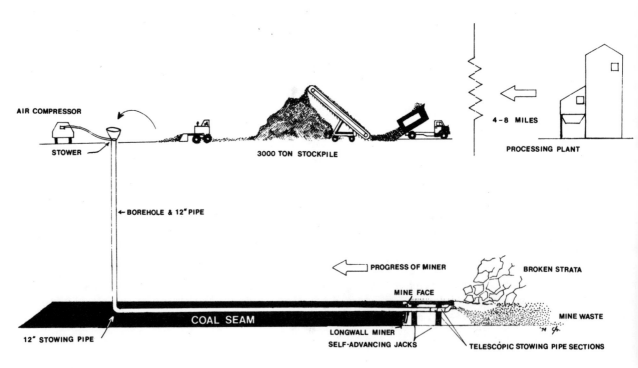

FIGURE 4.3   Underground Disposal Using Pneumatic Backfilling.

• It is assumed that maximum production from the longwall panel is 450 tons per hour. Although the stowing machine could backfill two cuts of the longwall at one time, the stowing system will have to keep up with the progress of the longwall face within two cuts if even backstowing of material is required for subsidence control or for maximum waste disposal. Longwall panels are capable of producing more than 450 tons per hour, but is assumed that peak stowing capacity is 450 tons per hour. If backfilling is strictly for waste disposal and the amount of material to be disposed is considerably less volume than the volume of coal removed, the peak capacity of the system ceases to be a problem (Table 4.5c).

• It is assumed that there are 230 working days per year with three eight-hour shifts per day. The actual working time at the face is approximately 6 1/2 hours, because approximately one hour per shift is spent transporting the mining crews to and from the working face and one half hour is allocated for dinner. This is typical work time for U.S. mines. It is assumed that it takes 20 days to move and set up the mining equipment on a new longwall panel.

• The coal seam that is being mined is assumed to be 5-feet thick, flat lying, and 400 feet below the surface. Coal is mined by longwall methods in seams as thin as 26 inches in Europe. In U.S. mines seams as thin as 28 inches and as thick as 20 feet are mined by room and pillar methods, and as thin as 40 inches and as thick as 7 feet by longwall methods. A 5-foot seam would provide excellent conditions for backfilling. That the seam is flat lying does not affect the facility of backfilling, but the depth of over-burden affects the cost of boreholes. A mine 400 feet below the surface is a relatively shallow mine.

• It is assumed that the mine is in a sparsely populated area with low real estate values, but serviced by an existing road network. In any disposal area there will be some problems associated with neighbors disgruntled by the noise of trucking or unwilling to sell or lease their land for temporary disposal areas. The cost of such inconveniences to the mine operator is not included in the committee's cost estimate. The longwall panel designated to be the waste area for the mine is assumed to be 3 or 4 miles from the preparation plant. This is probably an underestimate as most operating faces are at least four to eight miles from the preparation plant.

• The longwall panel dimensions are assumed to be 450 feet x 3000 feet in plan dimensions. Longwall panels vary in size and this would be a typical panel (see Chapter 3).

• Wages are computed on existing wage scales and include fringe benefits, payroll taxes, workmen's compensation and administrative costs. No production royalty has been added to the base rate for the United Mine Workers health and welfare fund. A breakdown of the wage scale is presented in Appendix C, Table B.

TABLE 4.5a   Committee's Estimate of the Cost of Pneumatic Backfilling Behind a Longwall Face in an Equilibrium Situation Where the Waste Produced by the Entire Mine Can Be Accommodated in a Single Mined-out Longwall Panel

Operation	Total Amount or Number	Fixed Costs Total/ yr.	Cost/Ton Clean Coal	Costs Related To Coal Production Total/ yr.	Cost/Ton Clean Coal	Costs Related To Refuse Handled Total/ yr.	Cost/Ton Clean Coal	Totals Total/ yr.	Cost/Ton Clean Coal
Crushing & Loading									
Capital	$198,800	$ 51,400	$.028						
Labor	1 man/shift			$ 45,500	$.025				
Oper., supply & power						$ 5,600	$.003		
TOTAL								$102,500	$.056
Surface Transportation									
Contract Trucking						$364,000	$.197	$364,000	$.197
Storage & Handling									
At the plant									
capital	$ 53,500	$ 13,800	$.007						
labor	1 man-shift			$ 15,000	$.008				
oper., supply & power						$ 5,600	$.003		
contract trucking						$ 23,000	$.012		
At the borehole									
capital	$265,000	$ 68,600	$.037						
labor	3 men-shifts			$ 45,500	$.025				
oper., supply & power						$ 43,000	$.023		
TOTAL								$214,500	$.115

TABLE 4.5a (Continued)

Item	Quantity								Total	
Boreholes & borehole sites										
Capital	400-ft boreholes		$ 70,000	$.038						
TOTAL									$ 70,000	$.038
Stowing system										
Capital		$711,000	$184,000	$.099						
Labor	11 man-shifts				$167,000	$.090				
TOTAL							$600,000	$.324	$951,000	$.513
Disposal of fines										
Capital		$500,000	$129,400	$.070						
Labor	50 man days				$ 3,300	$.002				
Maintenance							$ 2,800	$.002		
TOTAL									$135,500	$.074
Subtotal		$517,200	$517,200	$.279	$276,300	$.150	$1,044,000	$.564	$1,837,500	$.93
Productivity loss at the longwall										
Loss due to breakdown or plugging - 10%									$ .17	
Loss due to additional moving - .5%									$ .008	
Loss due to extending pipe - .5%									$ .009	
TOTAL									$1.180	

NOTE: A further breakdown of costs is presented in Appendix C, Table E--1,000 tons of refuse backfilled per shift; 2,575 tons per day waste produced from the entire mine; 1,850,000 tons per year cleaned coal from the entire mine.

TABLE 4.5b Committee's Estimate of the Cost of Pneumatic Backfilling Behind Two Longwall Faces When the Amount of Waste Produced by the Mine Exceeds the Capacity of a Single Longwall Panel

Operation	Total Amount or Number	Fixed Costs Total/ yr.	Cost/Ton Clean Coal	Costs Related To Coal Production Total/ yr.	Cost/Ton Clean Coal	Costs Related To Tons Refuse Handled Total/ yr.	Cost/Ton Clean Coal	Totals Total/ yr.	Cost/Ton Clean Coal
Crushing & Loading									
Capital	$198,800								
Labor	1 man/shift								
Oper., supply									
& maintenance									
TOTAL		$ 51,400	$.020	$ 45,500	$.018	$ 7,700	$.003	$104,600	$.041
Surface Transportation									
Contract trucking						$498,800	$.197	$498,000	$.197
TOTAL									
Storage & Handling									
At the plant									
capital	$ 53,500								
labor	1 man/shift								
oper., supply									
& maintenance									
TOTAL		$ 13,800	$.005	$ 15,200	$.006	$ 4,700	$.002		
At the borehole									
capital	$423,000								
labor	2 men shift								
oper., supply									
& maintenance									
TOTAL		$109,000	$.043	$ 91,000	$.036	$ 67,500	$.027	$301,200	$.119

TABLE 4.5b

Item		Capital	Labor	Oper., supply & maintenance	Total
Boreholes & borehole sites					
Capital	400-ft boreholes	$140,000 $.055			
TOTAL					$140,000 $.055
Pneumatic stowing system					
Capital	$1,421,274	$368,000 $.145			
Labor	16 men 1 day		$243,000 $.096		
Oper., supply & maintenance				$1,200,000 $.474	
TOTAL					$1,811,000 $.715
Disposal of fines					
Capital	$500,000	$129,000 $.051			
Labor	50 man days		$3,300 $.001		
Oper., supply & maintenance				$3,800 $.002	
TOTAL					$136,100 $.054
Subtotal		$811,200 $.319	$398,000 $.157	$1,781,700 $.705	$2,990,900 $1.18
Productivity loss at the longwall					
Loss due to breakdown or plugging - 10%					$ .170
Loss due to additional moving - .5%					$ .008
Loss due to extending pipe - .5%					$ .009
TOTAL					$1.367

NOTE: A further breakdown of costs is presented in Appendix C, Table E--1,111 tons of waste backfilled per shift; 3,666 tons per day waste produced from the entire mine (333 tons of fines); 2,530,000 tons per year cleaned coal from the entire mine.

TABLE 4.5c Committee's Estimate of the Cost of Pneumatically Backfilling Behind a Longwall Face, the Waste Produced by the Longwall and Its Development Entries

Operation	Total Amount or Number	Fixed Costs Total/yr.	Cost/Ton Clean Coal	Costs Related To Coal Production Total/yr.	Cost/Ton Clean Coal	Costs Related To Tons of Waste Handled Total/yr.	Cost/Ton Clean Coal	Totals Total/yr.	Cost/Ton Clean Coal
Crushing & Loading									
Capital	$198,000	$ 51,400	$.068						
Labor	1 man/shift								
Oper., supply & maintenance				$ 45,500	$.060	$ 2,300	$.003		
TOTAL								$ 99,200	$.131
Surface Transportation									
Contract trucking						$150,000	$.197		
TOTAL								$ 150,000	$.197
Storage & Handling									
At the Plant									
Contract trucking & loading						$ 9,700	$.013		
At the borehole sites									
capital	$118,500	$ 30,700	$.040						
labor	1 man/shift								
oper., supply & maintenance				$ 45,500	$.060	$ 24,600	$.032		
TOTAL								$ 110,500	$.132

TABLE 4.5c   (Continued)

Item	Capital	Amount	$/ton	Amount	$/ton	Amount	$/ton	Total	$/ton
Boreholes & borehole sites									
Capital									
400-ft boreholes								$ 70,000	$.092
TOTAL								$ 70,000	$.092
Pneumatic stowing system									
Capital	$705,000	$183,000	$.241						
Labor — 7 men/day				$106,000	$.211				
Oper., supply, & maintenance						$450,000	$.592		
TOTAL								$ 739,000	$1.044
Disposal of fines									
Capital	$500,000	$129,000	$.170						
Labor — 50 man days				$ 3,300	$.004				
Oper., supply, & maintenance						$ 1,200	$.002		
TOTAL								$ 133,500	$ .176
Subtotal		$464,100	$.611	$200,300	$.335	$637,800	$.839	$1,302,200	$1.772
Productivity loss at the longwall									
5% loss									$ .085
TOTAL									$1.857

NOTE: A further breakdown of costs is presented in Appendix C, Table E--333 tons of waste back-filled per shift; 1,100 tons per day waste production from the entire mine (100 tons of fines); 760,000 tons per year clean coal from the mine.

● A capital recovery factor that takes into account a 15-percent return on investment, a depletion allowance, depreciation over 10 years, and federal and state income taxes is charged against all capital expenditures to compute the investment per year for equipment and land (see Appendix C, Table C).

● It is assumed that the fines of the preparation cycle have to be impounded on the surface during pneumatic disposal just as with surface disposal methods, because the fine slurry cannot be transported by truck until it has been dewatered.

● A pneumatic system can handle material up to three inches in size. A crusher would have to be installed at the preparation plant to reduce the size of the oversized waste. Crushing all the waste with a jaw crusher or hammer crusher appears to be less expensive than sorting out the oversized pieces and crushing them separately.

● It is assumed that there will be a loss of mine productivity due to backfilling. With the addition of another system (i.e., the stowing machine) to the existing extraction system the downtime of the backfilling apparatus will result in downtime for the entire system and a production loss for the entire mine. This is particularly true in the equilibrium situation where the amount of waste to be disposed equals the volume of coal extracted and the pace of extraction must equal the pace of backfilling (Table 4.5a). Some of the downtime of the stowing system will coincide with downtime of the longwall mining equipment or transport system, however, some of the downtime will not coincide. A ten minute production loss results in a one percent loss of production per day. Production loss due to additional downtime of the backfilling system as a result of breakdown or plugging is estimated to be 10 percent and for extension of the pipe and additional moving, 1 percent.

As discussed in Chapter 3, there may be problems associated with backfilling whose magnitude have not been assessed due to lack of experience in U.S. mines. Some of the potential problems which have not been assessed or resolved include problems of noise, dust, roof control and rebounding rock associated with pneumatic backfilling. These problems may be solved if enough time and money is expended on them, but estimates of these costs were not included in the committee's analysis of the costs of pneumatic backfilling.

4.6 COMMITTEE ESTIMATE OF HYDRAULIC BACKFILLING

In designing a hypothetical hydraulic backfilling system the committee tried to envisage a mine with representative conditions that would result in the most economically feasible system. Hydraulic backfilling has been used extensively to backfill abandoned mines for subsidence control (see section 4.3) and is used in Poland and other European countries to backfill in operating mines. Hydraulic backfilling is most practical in active U.S. coal mines containing partially mined areas that can be used as disposal areas rather than in the active operating areas of the mine. It can only be used where a mine is sufficiently developed to have the mined out panels to be backfilled or where a preparation plant is located near abandoned workings. The committee's estimate of costs is projected for a mine similar in size and geologic conditions to that of the equilibrium case for pneumatic backfilling (see section 4.5). The system is shown in Figure 4.4.

Because the hydraulic disposal system is operated independently from the extraction system, it is more flexible in terms of layout and operation than the pneumatic system envisaged in section 4.4. However, the geologic conditions of the disposal area take on greater importance because the transport medium is water, a significant problem in mines. The specific assumptions used in the committee's projection are summarized below. Costs are presented in Table 4.6 and described more fully in Appendix C, Table E.

As with the pneumatic stowing system, waste is assumed to be 25 percent of raw coal production and to occupy the volume of the waste void created by an equal tonnage of raw coal. It is assumed that a single partially mined area is set aside to accommodate the waste from the entire mine. Because the disposal site is independent of the extraction operation, the amount of material to be disposed per day does not affect the basic design of the system, only the size and capacity of the equipment. For fair comparison, it is assumed that 1,000 tons of waste are stowed each shift. Such a mine would produce 2,070,000 tons per year of clean coal.

Hydraulic systems can easily accommodate the peak quantity of waste produced from the preparation plant (see Chapter 3). The system projected in this report would be designed to backfill 200 tons of refuse per hour at peak capacity.

- It is assumed that the backfilling operates 230 working days per year with three 8-hour shifts. Unlike the pneumatic system which does not operate while the longwall is being moved, there is no major loss in time in moving the hydraulic system from one borehole site to another because the surface operation is set up on a truck and therefore is mobile.

- It is assumed that the coal seam is 5-feet thick gently dipping and approximately 400 feet below the surface. In order to minimize the seepage of water to other sections of the mine the disposal area is in naturally low areas, i.e., synclines or at the low points of pitching seams. It is assumed for this model that there are naturally occurring sumps in the disposal area or adjacent to it for collection and recirculation of water.

- It is assumed that the dimensions of the partially mined out panel measure 500 feet x 2000 feet x 5 feet, i.e., a volume of 275,000 cubic yards. Because the panel is mined out leaving 50 percent of the coal in place, only 138,000 cubic yards of void remain of which only 75 percent of the volume would be backfilled. This void volume of 100,000 cubic yards would accommodate 100,000 tons of refuse. Three boreholes per panel should be sufficient to inject this amount of waste material. This would amount to 7.5 panels (500 feet x 2000 feet x 5 feet) each year to accommodate 690,000 tons of refuse.

- Of primary concern in the hydraulic backfill system is the prevention of water seepage to the active portion of the mine. For this reason it is assumed that as much water as possible is recirculated in the system. In

TABLE 4.6 Committee Estimate of the Cost of Hydraulic Backfilling

Operation	Total Amount or Number	Fixed Costs		Costs Related To Coal Production		Costs Related To Refuse Handled		Totals	
		Total/ yr.	Cost/Ton Clean Coal	Total/ yr.	Cost/Ton Clean Coal	Total/ yr.	Cost/Ton Clean Coal	Total/ yr.	Cost/Ton Clean Coal
Crushing & Loading									
Capital	$399,000	$103,000	$.050						
Labor	3 men/day								
Oper., supply, & power									
TOTAL				$ 45,500	$.022	$ 77,300	$.037	$ 225,800	$ .109
Surface Transportation									
Trucking	Contract					$448,500	$.217	$ 448,500	$ .217
Borehole & Site									
Capital	$267,000	$267,000	$.129						
Labor	1 man/day								
Oper., supply & power									
TOTAL				$ 15,200	$.007	$ 3,000	$.001	$ 285,200	$ .138
Stowing Unit & Surface Water Supply									
Capital	$530,000	$326,000	$.157						
Labor	3 men/day								
Oper., supply & power									
TOTAL				$ 45,500	$.021	$ 31,000	$.015	$ 402,500	$ .194

TABLE 4.6 (Continued)

Underground Pumping & Recirculation						
Capital	$227,600	$85,300 $.041				
Labor	2 men/day		$ 30,400 $.015		$ 20,000 $.010	
TOTAL						$ 135,700 $ .066
Underground Storage Site						
Capital	$ 7,800	$ 7,800 $.004				
Labor			$ 59,600 $.029			
TOTAL						$ 67,400 $ .033
Subtotal		$789,100 $.381	$196,200 $.094		$579,800 $.280	$1,565,100 $ .757
Productivity loss to the entire mine						
10% loss ($.067 loss per 1% loss)						$ .670
TOTAL						$1.425

NOTE: A further breakdown of costs is presented in Appendix C, Table F--2,070,000 tons of clean coal per year; 690,000 tons of refuse produced per year; 1,000 tons of refuse backfilled per shift.

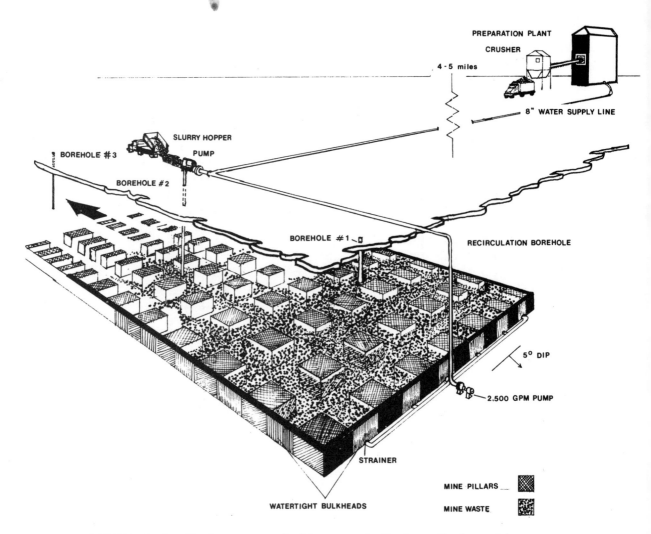

FIGURE 4.4 Underground Disposal Using Hydraulic Backfilling.

order to do this, bulkheads would be set up at the margins of the partially mined panel being backfilled and a sump designed into the system. It is assumed that 70 percent of the slurry water is recirculated, but that 30 percent of the water is lost through seepage and would flow to mine sumps throughout the mine and have to be pumped from those areas.

• It is assumed that water supplied to the circulation system consists of water recirculating in the system supplemented by water from the preparation plant or from the other mine sumps.

• In order to be carried in suspension for over 100 yards waste material must be of smaller size than that transported pneumatically. It is assumed that minus 3/4 inch particle size, the maximum size used in hydraulic flushing projects, is sufficiently small.

● The committee assumes surface transport of waste by truck from the preparation plant to the borehole site. This system is similar to that of the pneumatic system. Although under propitious circumstances, hydraulic transport of the waste material from the preparation plant to the disposal site would be feasible, it would necessitate four times the amount of water to transport the material horizontally than simply to inject it. Such quantities of water could increase seepage to the active portion of the mine, and lessen productivity, increase the cost of recirculation of water, and necessitate extensive bulkhead sealing. It would be necessary to install at considerable expense a system to partially dewater the slurry at the injection site, and a recirculation system to return the excess water to the plant for reuse. For these reasons, the committee preferred a model that used surface transport of the wastes from the preparation plant to the borehole site.

● Water seeping through the coal seam and adjacent rock strata could cause a production loss to the entire mine. Shuttle cars could become stuck and rail and haulageways deteriorate more rapidly than normal. It is conservatively estimated that the loss of productivity due to such seepage would be 10 percent. This item is the single greatest cost to hydraulic backfilling.

● Hydraulic backfilling could result in a greater quantity of water to be treated to reduce acidity before being returned to surface streams. The cost for such treatment is not included in the committee's estimate.

● The committee's assumptions regarding local land use pattern, wages, and opportunity cost of capital are the same as those for pneumatic stowing (see section 4.5).

CHAPTER 5

THE LEGAL PARAMETERS AFFECTING DISPOSAL OF COAL MINE WASTES

SUMMARY

The federal regulations governing the surface disposal of coal were promulgated in 1971 pursuant to the Federal Coal Mine Health and Safety Act of 1969. Amendments to these regulations have been suggested to strengthen their requirements and fulfill the intent of the laws.

In the aftermath of the Buffalo Creek disaster in 1972, laws were passed to require an inspection and inventory of all waste impoundments in order to determine their stability. As a result of these inspections, many impoundments were declared hazardous and their owners have been required to correct the situation. Legislation continues to be put forward in Congress to insure the safety of the workers and local inhabitants from the dangers of waste piles and impoundments.

Federal water pollution requirements and air pollution requirements affect waste disposal indirectly. Proper underground or proper surface disposal practices result in nonpolluting disposal sites.

State laws affecting the disposal of coal mine wastes vary considerably. Pennsylvania has a regulatory program to control subsidence from coal mining and has stricter safety, air pollution, water pollution and general waste disposal regulations than other Appalachian states such as West Virginia and Kentucky.

In spite of the legislation and regulations intended to insure environmentally acceptable disposal of mine waste, dangerous and polluting waste piles and impoundments constructed through the use of improper disposal practices are in evidence throughout the eastern coal fields.

5.1   FEDERAL REGULATIONS REGARDING DISPOSAL OF MINE WASTES

The Coal Mine Health and Safety Act of 1969

The Federal regulations which govern the surface disposal of coal wastes are promulgated pursuant to the Federal Coal Mine Health and Safety Act of

1969. Under section 3 (Definitions) the words "coal mine"* can be interpreted to include coal mine waste materials. Moreover, Section 101(h), Title 1 of the Act directed the Secretary of Interior to develop health and safety standards for surface work areas related to underground coal mines. Three regulations were published in the Federal Register on May 22, 1971 and amended on July 15, 1971. Further amendments are now being considered as published in the Federal Register on January 16, 1974.

The first of these regulations (section 77.214 of the Code of Federal Regulations) deals with the general question of waste piles, giving particular attention to new piles. It requires that waste piles constructed after 1971 be located a safe distance from underground mine openings and preparation plants and that new refuse piles constructed over exposed coal beds are to be separated from the coal by clay or some other inert material.

Section 77.215 requires that certain standards be met when constructing coal waste piles. It requires that the waste be spread in layers and placed in such a manner as to minimize the passage of air through the piles, but does not specify the thickness of the layers, whether the waste must be compacted or the steepest slope or the height to which a waste pile may be constructed. Refuse piles are not to be constructed in a way that impedes drainage or impounds water and requires that the waste piles be constructed in such a manner as to prevent accidental sliding and shifting of materials. No combustible material can be placed on a waste pile according to this regulation.

Section 77.216 specifies construction, inspection and monitoring of all retaining dams. If failure of a water or silt-retaining dam would create a hazard to miners, it should be "of substantial construction" and be inspected at least once each week by a certified person who is required to report the inspection and have the report countersigned by a foreman or superintendent of the company.

The Coal Mine Safety Inspection Manual which clarifies these regulations for practical application by inspectors explains that a safe distance as required in the Section 77.214 means at least 300 feet away from all underground mines or shafts, preparation plants, tipples, etc. It explains the requirement of Section 77.215 that with the use of earth grading equipment the waste material is to be spread in layers not to exceed two feet thick and compacted with compaction machinery or heavy equipment. It also explains that "substantial construction" as referred to in Section 77.216 means that the structure is "solid, firm, or stable, and has been constructed according to sound engineering principles recognized for construction of water retaining structures."

---

*"...any area of land and all structures, facilities, machinery, tools, equipment, shafts and slopes, tunnels, excavations, and other property, real or personal, placed upon, under, or above the surface of such land by any person, used in or to be used in or resulting from the work of excavating in such area bituminous coal, lignite, or anthracite, from its natural deposits in the earth by any means or method, and the work of preparing the coal so extracted, and includes custom coal preparation facilities."

From the point of view of the coal mine operator or inspector these explanations lack practical and enforceable specificity. Federal inspectors from the Department of the Interior's Mining Enforcement and Safety Administration (MESA) have described how such inspections take place. An inspector during the normal course of his duties will look at the waste piles and water impoundments associated with a mine when he makes an inspection of the mine. Should he see any sign of instability, he will then report it to his supervisor, generally the District Manager. The District Manager then institutes a more detailed inspection of the waste pile or impoundment if he feels the pile is unstable or if it is in violation of Section 77.216. This more comprehensive inspection is done by MESA's technical team of experts who are trained in coal waste pile stability. If, as a result of this investigation, the waste pile is found unstable, a detailed engineering study is ordered. This engineering effort is to develop standards and criteria that will make the specific waste pile stable. All waste pile cases are handled on an individual basis.

According to MESA records for 1973, there were approximately 90,000 routine inspections of approximately 1,692 surface mines and 1,425 underground mines. In addition to these generalized inspections, all of 652 surface structure inspections were for the sole purpose of inspecting surface disposal sites of underground mines particularly waste impoundments (waste piles that impound or that are capable of impounding water) rather than "dry piles." These inspections led to inspections of 175 sites by the team of experts from MESA and the Bureau of Reclamation. Records of MESA do not specifically indicate how many sites were considered unstable by a) the MESA inspector, b) the District Manager, c) the team of experts, or d) by consulting engineers, nor is it easy to ascertain how many piles or impoundments have been altered to meet MESA specifications because of these inspections.

In addition to its routine site inspection and regulatory program, MESA is upgrading and expanding its activities concerning coal mine waste piles and impoundments. These efforts will consist of completing the emergency investigation of coal mine waste embankments initiated after Buffalo Creek, studying the physical properties of waste, issuing a manual for inspectors to aid their inspection procedures and set up a new hazard rating system, writing an engineering design manual for constructing surface disposal sites, and instituting a training program for inspectors to improve their ability to evaluate, recognize and report basic unsafe conditions at waste piles and impoundments.

The emergency investigation of 139 impoundments resulted in the rating of 8 as imminent hazards, 66 as obviously deficient, 61 as not obviously deficient, and 4 as missing data. Some action was taken on all 8 imminently hazardous sites. Three conclusions of the interim emergency report were as follows:

A. Coal mine waste impoundments can and do constitute potential hazards to miners, miner property, and the general public.

B. Coal mine waste embankments that do not impound water and sludge can also be hazardous.

C. Presently available and immediately foreseeable technology cannot be expected to prevent all future failures. New or evolving technology is

currently needed to significantly reduce the probability, severity and consequences of failures. (U.S. Department of the Interior, 1972a, p.92--Final
report to be issued in 1974.)

In another study completed in 1972, 960 significant waste disposal piles were
inspected of which 670 were dry piles, 290 were impounding water, and 557 were
burning or impounding water (Department of the Interior, 1972).

The federal regulations concerning coal mine waste disposal that were
promulgated pursuant to the Coal Mine Health and Safety Act of 1969 were in
effect at the time of Buffalo Creek in 1972 and have not been changed since
then. In January, 1974 the Department of Interior published in the Federal
Register proposed new and expanded regulations controlling waste piles. These
regulations are meant to provide more detailed guidance to the operator and
inspector, and to be more comprehensive than the 1971 regulations. In an
environmental impact analysis issued with the proposed regulations, MESA
justified its actions.

> Evaluations of more than 200 coal waste disposal sites, made by
> State and Federal agencies and private consultants, have revealed that
> the majority of those sites evaluated were judged to be potentially
> hazardous from a stability standpoint.* In nearly all cases, corrective
> measures were necessary before the stability of the deposit could be
> assured. In view of the fact that the methods used to deposit the coal
> waste material at these sites not investigated varied little, if any,
> from the methods used at those sites which were evaluated, it must be
> assumed that the majority of all coal waste deposits are potentially
> hazardous.

The proposed regulations would require (1) a certification of stability by
the operator and a certified engineer that "the refuse pile cannot impound
water and is constructed in such a manner as to preclude the probability of
failure of such magnitude as to endanger the lives of coal miners," (2)
identification and registration of the waste piles, and (3) an engineering
analysis of the waste pile's stability, capacity for impounding water, potential hazards and environmental effects (runoff characteristics, combustion
potentials). The regulations would establish a limit to the amount of water
allowed to be impounded behind each pile. The proposed regulations for waste
piles are less stringent than those for waste impoundments. The maximum slope
for a waste pile would be less than 27° and the operator would be required to
compact the waste in layers less than 2 feet thick. The new regulations also
would require the operator to make weekly inspections on a more stringent engineering basis than now and to keep his own records on inspection, construction techniques and material content of each waste pile.

---

*Records on file in the Mining Enforcement and Safety Administration's Offices
of Technical Support and Coal Mine Health and Safety of Investigations by
MESA; the Department of the Army, Corps of Engineers; Bureau of Reclamation;
and various consulting firms. (U.S. Department of the Interior, 1973b, p.5.)

The Federal laws directly applicable to the stability of waste piles and waste embankments have changed little since the Buffalo Creek disaster. Although enforcement of these laws has been increased, the requirements and regulations adopted under these laws have not changed substantially. Where impoundments are found to constitute danger to miners, mine operators will be required to place spillways and emergency drainage ditches and to take other precautionary measures. They are explicitly required to build new waste piles no less than 300 feet from the mine, not to have these piles impede drainage or impound water and finally, to make sure that these waste piles are of substantial construction. But some old active piles remain which impede drainage and impound water, are not of substantial construction, constitute potential hazards to the individuals working in mines and persons living nearby and cause air and water pollution. The laws and regulations virtually ignore the question of abandoned and orphaned waste piles.

The Dam Safety Act of 1972

In the wake of the Buffalo Creek disaster, Public Law 92-367 was enacted on August 8, 1972. It required the Secretary of the Army through the U.S. Corps of Engineers to conduct a survey of all mine waste impoundments in the Appalachian region and report to the governors of the respective states and to Congress before July 11, 1974. The report is to include an inventory of the impoundments, a review of inspections, recommendations for an inspection program, a suggested delineation of responsibilities for waste impoundments, and a national program relative to mine waste embankments. It is expected to be completed during the summer of 1974. All coal mine waste impoundments in Appalachia are being inspected by a technical team consisting of three to five technical people from the Corps of Engineers, who ascertain the relative stability of each site and also the potential damage it could cause should it fail. Periodically a review board considers the recommendations of the technical team and informs the operator, the state agency, and MESA of its conclusions. The Corps of Engineers has no authority to take corrective actions, but when an impoundment presents an imminent hazard, the responsible state agency and MESA as well as the operator are immediately informed. The Corps of Engineers is inspecting at least 1,000 active, inactive and abandoned waste impoundments. It appears that at least 60 percent of these impoundments probably do not present a hazard to the public. In all cases where imminent hazard has been determined, corrective action has been taken by the operator in compliance with MESA inspection. Only one operator has been threatened with a shutdown of operations.

Flood Control Act

The Corps of Engineers under Section 233 of Public Law 91-611 has the responsibility of investigating the stability of all medium-sized and larger water impoundments to determine potentially hazardous flooding conditions.

The inventory in some cases will be done by the states themselves and some are now completed. The final inventory will include many coal waste impoundments and to some extent overlaps the inventory prepared under the Dam Safety Act.

HR 11500

The most recent legislative discussion that considers coal waste piles is HR 11500, the Surface Mining Control and Reclamation Act of 1973, a bill in the House of Representatives which considers the surface effects of underground mining as well as those of surface mining. The bill as reported from subcommittee included a provision which required that, as far as technologically and economically possible, mine wastes and coal processing plant tailings from new operations be returned to mine voids or disposed of in other equally environmentally protective methods. The bill also specifies that the latest available engineering practices be used in the design and construction of water retention facilities in order to insure that such impoundments will not endanger the health and safety of the public. The bill also addresses the problems of abandoned or orphaned piles by assigning high priority to them in its directive to the Secretary of the Interior to insure the health and safety of the public. As of April, 1974, this bill had not become law.

Federal Water Pollution Requirements

The Federal Water Pollution Control Act as amended in 1972, PL 92-500, has set a goal of zero polluting discharges in 1985. To achieve this goal the Act requires the "best practical technology" to be implemented by 1983. The Act provides for a "National Pollutant Discharge Elimination System," requiring a permit for the discharge of any pollutant, or combination of pollutants. Mining companies must file for a discharge permit, but standards for discharge have not been set. New minimum standards will be promulgated by early 1975 and will be more stringent than the current interim guidelines. These standards will be revised for implementation in 1983 to the "best available technology." How these standards will affect coal waste disposal is unknown at this time.

Federal Air Pollution Requirements

The Clean Air Amendments of 1970 contain two essential provisions. First, they require the Environmental Protection Agency to establish federal ambient air quality standards for the entire country. Second, standards of performance for new stationary sources of air pollution are to be established. After the nationwide ambient air quality standards are established, each state which desires to conduct a state regulatory program must submit an implementation plan which provides systems for the state or region within the state to achieve the ambient air quality standards.

"Stationary sources" include any building structure, facility, or installation which emits or may emit any air pollutant. The term "new source" includes modifications of a stationary source. Thus, construction of a new waste refuse pile or expansion of an existing one would be considered a "new source."

However, at the present time coal waste piles are not included in lists of EPA's categories of stationary sources that contribute significantly to air pollution and no regulations are, therefore, being proposed at this time. One apparent reason for this is that a coal waste pile is not normally thought of as a source of air pollution as are chimneys and stacks and other air pollution sources associated with a manufacturing process. An obvious performance standard for a waste pile is no ignition, and therefore, no pollution.

Power generating stations that burn coal have generally found that an economic way to meet sulfur emission standards is to lower the sulfur content of the coal they buy from coal processing plants, should low-sulfur raw coal not be available. The processing of high-sulfur coal in plants to lower sulfur content will increase the percentage of coal waste generated from each ton of coal. Although the trend during the 1960's was toward burning run-of-the-mine coal, as enforcement of the sulfur emission standards becomes more strict, more sulfur will be removed by coal processing rather than by removing sulfur from stack emissions. This will increase the amount of coal waste generated and its potential for water and air pollution.

5.2   CURRENT STATE LAWS AFFECTING THE DISPOSAL OF COAL MINE WASTES

The laws of three states, Pennsylvania, West Virginia, and Kentucky, are considered in this report.

Pennsylvania

At the present time Pennsylvania has the most stringent regulations to control the surface disposal of coal waste. It is also the only state which has a regulatory program to control damage from subsidence and a state operated insurance program enabling homeowners and commercial establishments to acquire low-cost subsidence casualty insurance.

The state's first attempt to control coal waste disposal was primarily aimed at preventing the ignition of coal waste. Accordingly, this first effort was in the form of a regulation adopted in 1962 pursuant to the state's air pollution control act. Apparently this regulation was quite successful since it is reported that no piles created subsequent to 1962 which complied with the regulation have ignited.

In 1965, the state's Clean Streams Law was amended to require the abatement of water pollution from coal mines and associated activities such as coal waste piles. Prior to that time, pollution abatement from mining activities was specifically excluded in the Clean Streams Law, except where new sources of pollution from mines were proposed to be created which would discharge into "clean streams" (streams essentially free of acid mine drainage).

In 1968, primarily as a result of the Aberfan Disaster, the Pennsylvania General Assembly enacted the Coal Refuse Disposal Control Act which gave the state government the authority to abate safety hazards at coal waste piles deemed to be subject to an imminent danger of sliding or shifting. This act also spelled out enforcement steps that state inspectors could take under existing pollution control laws to correct air and water pollution from coal waste piles. The act did contain the weakness of *not* giving authority to require a permit system whereby plans for the creation of new coal waste piles would have to be approved prior to initiation of a coal processing operation. It, therefore, did not provide preventive powers with respect to safety hazards or pollution.

The lack of definitive technical requirements or a review and permit mechanism prior to creation of a new coal waste pile, together with the lack of requirements for proof of financial responsibility by coal operators, significantly handicapped state regulation of the waste piles. In order to correct this deficiency, regulations were adopted in May of 1973, which when fully implemented should provide strict regulation of coal waste disposal on the surface of the land.

Among other things the regulations require:

1. Operators of existing coal waste piles to apply for a permit within 6 months after adoption of the regulations.

2. Operators to obtain permits before creating new refuse piles.

3. The applicant to provide proof of his financial responsibility with such responsibility continuing for up to 10 years following the completion of the disposal operation to assure proper closure of sites.

4. The application to include maps and other technical information pertaining to soils, geologic and groundwater characteristics of the area and the possibility of subsidence from past and future mining beneath the disposal area. The plan must provide for the prevention of water pollution, the stability of the disposal area and the prevention of air pollution. Drainage from the coal waste areas must meet the water quality standards established for the particular receiving streams.

5. Approval for the construction of impoundments or silt lagoons on coal waste areas under a desparate permit subject to the state's existing regulations for design and construction of impoundments.

6. The side slopes of a waste pile not to exceed 33 percent (18°). A vegetation plan for the entire area is mandatory. All earth moving and re-vegetation activities must comply with the state's existing regulations for erosion and sediment control.

7. Concurrent compaction of the coal refuse to minimize the pile's permeability and air entraining capabilities.

The use of coal waste as a highway fill material is now approved in Pennsylvania and its use is rapidly accelerating, as the state's Department of Transportation is making a conscious effort to design its road construction so as to utilize the maximum amount of coal waste which can reasonably be included in a highway plan.

Subsidence Regulation

Increased public concern about damage to homes from subsidence resulted in the Pennsylvania legislature passing a mine subsidence regulation act in 1966. This act requires the protection of all dwellings, public buildings and cemeteries constructed or established prior to April, 1966, from any bituminous coal mining undertaken after that date. The owners of nonprotected surface structures are given the opportunity to buy sufficient coal to be left in place as support pillars to reasonably assure protection of the structure. If the nonprotected surface structure owner chooses not to purchase the coal, the mining company is not liable for damage. If damage should occur to protected structures despite the precaution of only partial mining to provide support pillars and just compensation for damages is not forwarded to the surface structure owner within six months, the mining company is required to place funds in escrow with the state to cover the amount of damage until the matter is settled.

In its almost eight years of operation, this program has had good success and seems to meet the needs of the people while not being repressive to the coal industry in Pennsylvania. More than 12,000 structures have been protected under this program and less than 1 percent have suffered damage.

As a further protection to surface structure owners in Pennsylvania, a subsidence insurance program is operated by the state government which provides low-cost protection particularly to those homes and other structures that are located over old deep-mined areas. At the present time a homeowner can acquire insurance for his home at an average rate of $3.50 per year per $1,000 of coverage. The law was recently amended to allow the coverage of commercial establishments and other structures. The premium cost for such coverage is slightly higher than that for a private home.

The subsidence insurance program has generally been successful, although the program administrators feel that the public has not taken full advantage of it. Increased efforts are underway to publicize the program's existence to encourage public participation.

West Virginia

The laws intended to regulate the disposal of coal waste in West Virginia are administered by the state's Department of Natural Resources. Subsequent to the Buffalo Creek disaster of 1972, the West Virginia Legislature passed two acts governing the disposal of coal waste. The first in 1972, entitled Coal Refuse Disposal Control Act, contained legislative findings expressing the intent to protect public life and property. It directed that within one year the Department of Natural Resources was to inspect and to make an inventory of all coal waste disposal piles and associated water impoundments and on the basis of competent engineering studies evaluate the stability of these waste piles and whether such piles would remain stable if subjected to climatological conditions resulting from a 50-year flood. The Director also was empowered to take remedial action when he found an imminent danger to human life. He could secure such piles in order to abate the conditions which caused the danger or apply for injunctive relief when he found a dangerous condition with respect to any *new* coal waste disposal pile created after 1972 or any part of a coal waste pile of an active operation. The Director's authority enabled him to require the operator to take all remedial actions which might be necessary to prevent or correct the condition. The operator who was thus affected had the right to a hearing before the Director to appeal such a decision.

The study by the Department of Natural Resources was completed in 1973 in conjunction with the Army Corps of Engineers. Of 531 significant coal waste impoundments 117 were considered hazardous, 101 possibly hazardous, and 363 probably not hazardous. Remedial actions have begun on many of the more hazardous impoundments. No operator has been cited for noncompliance with the regulations although several water impoundments have been discontinued and drained.

The second act, known as the Dam Control Act of 1973, broadened the power of the Department of Natural Resources so as to "...provide for the regulation and supervision of dams...to the extent necessary to protect the public health, safety and welfare. The act empowered the Director of the department to control and exercise regulatory jurisdiction over dams in the state. He was to review all applications for certificates of approval for the placement, construction, enlargement, alteration, repair or removal of any dam. Upon review of such applications, he was empowered to grant and revoke, restrict, or refuse to grant any certificate of approval based upon a determination by him that such action is proper or necessary to protect life and property as provided in this Act. The Director was also given power to: (1) adopt and repeal rules and orders which would allow him to implement and make effective his powers; (2) take lawful action to enforce the provisions of this article; (3) establish a charge of not more than $25 to review applications of approval for such dams; (4) employ qualified consultants and additional persons as necessary to review such applications; (5) coordinate with agencies of the federal and county governments; (6) improve and study dam safety; (7) make all investigations or inspections necessary to implement or enforce provisions of this article at any time as may be necessary to do so; and (8) prepare and publish criteria covering the design, construction, and maintenance of the proposed dams.

The criteria makes it unlawful to construct or enlarge such impoundments without a certificate of approval and it required that plans and specifications for placement, construction, and enlargement be in the charge of a registered professional engineer. The right to a hearing for the application for certification is granted to the operator and any person whose life or property could be adversely affected by the construction or existence of such an impoundment. The certificates of approval can be revoked or suspended if its requirements are not met by the operator. Should the Director feel circumstances have changed substantially, he may withdraw the certificate or amend its terms or conditions. The progress of work on any impoundment during its construction, enlargement, repair or removal is to be inspected. The owner of any impoundment has the primary responsibility of notifying the Director and any persons who may be in danger if the dam should fail and to take immediate remedial action to protect life and property as does the

Director of the Department of Natural Resources. An emergency telephone system has been instituted for the reporting of any knowledge or fear of a dangerous situation from coal mine waste disposal facilities. The Director also has the authority to issue and enforce orders for the correction of potentially hazardous conditions, e.g., he can take charge of the dam, lower the water level, drain all water from it, and perform any other remedial or protective actions at the site in order to safeguard life and property. The cost of such actions is to be paid by the Department of Natural Resources but reimbursed by the owner. Unless the Director determines that the dam is unsafe, impoundments that were completed in 1972 need not meet the criteria established by the Director although impoundments must be certified. The impoundments with any actual or potential problem are now routinely inspected.

Violation of the act is a misdemeanor and results in a fine not less than $100 and no more than $1,000 or imprisonment for no more than six months, or both. No violations have been reported.

Kentucky

The responsibility for safety of all dams is vested in the Department for Natural Resources and Environmental Protection. Legislation is now pending to give the Department additional authority over all coal refuse piles to regulate both stability and environmental aspects. Although the authority for requiring approval of all dams, including those constructed of coal refuse, has existed in Chapter 151 of Kentucky's Revised Statistics since 1967, lack of staff and enforcement have resulted in minimum compliance until the past year. The foregoing statute does not require approval of structural plans for dam construction and issuance of a permit. The stated purpose is to protect against threats to life and property of the public.

The state inventory of existing dams is nearly complete. Compliance action is also underway to correct coal refuse pile and refuse impoundment dam hazards as determined by the U.S. Army Corps of Engineers.

5.3 EUROPEAN REGULATIONS CONCERNING THE DISPOSAL OF COAL MINE WASTES

Having discussed U.S. regulations concerning coal mine waste disposal it is instructive to examine corresponding regulations in some European countries, where similar problems exist. As already mentioned, although underground disposal of the waste has been carried out in the past, its use has declined significantly in recent years. This implies a greater commitment to surface disposal; and with it has come a greater concern that surface disposal be carried out in ways that are safe and environmentally acceptable. The Aberfan Disaster in Wales in 1966 also stimulated attention to the potential hazards of surface disposal sites and has led to special new regulations [Mines and Quarries (Tips) Regulations 1971]. It is interesting to note that the tribunal appointed to inquire into the Aberfan disaster considered the question of underground disposal, but did not recommend this alternative. The comments of the tribunal on this point are contained in Appendix D along with a translation of an extract from regulations governing German coal mining in the Ruhr area. (A copy of the complete British regulations of August 17, 1971, Statutory Instruments No. 1377, can be obtained from Her Majesty's Stationery Office, Holborn Viaduct, London E.C. 1.)

Both the U.S. and European regulations are, of course, intended to accomplish the same ends of safe and environmentally acceptable disposal. In general, however, European regulations are specified in significantly greater detail.

Before a permit can be granted, for example, it is necessary to supply maps, cross-sections, etc., of the proposed method of disposal site construction, and information concerning the geological foundation upon which it is proposed to construct, details of groundwater regime and possible influence of the waste pile or impoundment on it and on surface water. Composition of the waste must be specified and the waste pile must be oriented with respect to prevailing wind direction so that the possibility of ignition of the waste material is minimized. According to the German regulations, the mine operator must demonstrate that the public interest as well as his own has been taken into account in his decisions of disposal of wastes, and that underground disposal or disposal in surface pits or mines would not be a more desirable alternative than surface disposal.

REFERENCES CITED

American Institute of Mining, Metallurgical, and Petroleum Engineers, Inc., 1968, *Coal Preparation*, J.W. Leonard and D.R. Mitchell (editors), (New York: AIME).

_____, 1973. *Elements of Practical Coal Mining*, S.M. Cassidy (editor), (New York: AIME).

The Appalachian Regional Commission, 1969. *Acid Mine Drainage in Appalachia*, (Washington, D.C.: ARC) 6v.

Appalachian Research and Defense Fund, Inc. 1972. Disposing of the Coal Waste Disposal Problem, Charlestown, West Virginia, 99 pp.

Ashcraft, J., personal communication, 1973-74. Director of the Department of Mines of West Virginia.

Baker, Michael, Jr., 1973. Analysis of pollution control costs, report to the Appalachian Regional Commission, ARC Contract No. 72-87/RPC-713, 436 pp.

Ball, D.G., personal communication, 1973-74. General Manager, Radmark Engineering, Ltd.

Ball, D.G. and J.E. Powell, 1972. High capacity pneumatic conveyor transport in underground mining, *Western Miner*, April edition.

Bishop, A.O., 1973. The stability of tips and spoil heaps, *The Quarterly Journal of Engineering Geology*, v.6, nos. 3 and 4, pp. 335-376.

Brauner, G., 1973. Subsidence due to underground mining: 1. Theory and practices in predicting surface deformation, 2. Ground movements and mining damage, *U.S. Bureau of Mines Information Circulars 8571 and 8572*, 56 and 53 pp.

Brieden, (Mashienenfabrick Karl Brieden & Company), personal communications, 1973-74, and various articles including: 1973. Movable Brieden blast stowage lines with lateral discharge units.

British Department of the Environment, 1972. Use of waste material for road fill and Appendix, *Circular 22/72 Welsh Office*.

British National Coal Board, 1963. Principles of subsidence engineering, *Production Department Information Bulletin 63/240*, 27 pp.

_____, 1966. Subsidence engineers' handbook, NCB Production Department.

_____, 1970. Spoil Heaps and Lagoons, NCB Technical Handbood, 232 pp.

_____, 1971. Tips, *NCB (Production) Codes and Rules First Draft*, 82 pp.

_____, 1972. Reviews of research on properties of spoil tip materials, *NCB Headquarters Research Project No. S/7307.*

_____, personal communication, 1973 and several articles including: Longwall full-face pneumatic stowing, 2 pp.

Candeub, Fleissig, and Associates, 1971. Demonstration of a technique for limiting the subsidence of land over abandoned mines, *National Technical Information Service,* PB-212 708, 87 pp.

Code of Federal Regulations, Sections 77.214, 77.215, 77.216, 30CFR77.

Cortis, S.E., 1968. Presentation to the roof control committee meeting of the American Mining Congress, Pittsburgh, Pennsylvania, September 18, 1968.

Cortis, S.E., 1969. Coal mining and protection of surface structures are compatible, *Mining Congress Journal,* v.55, no. 6, pp. 84-89.

Courtney, W.J. and M.M. Singh, 1972. Feasibility of pneumatic stowing for ground control in coal mines, *IIT Research Institute Report No. D6068 for U.S. Bureau of Mines,* 128 pp.

Crouch, S.L. and C. Fairhurst, 1973. Analysis of rock mass deformations due to excavations, In: D.L. Sikarski (editor), *Rock Mechanics Symposium,* (New York: AIME) pp. 25-40.

Dahl, H.D., 1972. Two and three dimensional elastic-elastoplastic analyses of mine subsidence, *Fifth International Strata Control Conference,* London, paper 28.

Dowell, division of the Dow Chemical Company, personal communication, 1973-74, and various articles including 1973. A summary of the Dowell closed-system hydraulic backfill process conducted in the Green Ridge Section of Scranton, Pennsylvania, for the Department of the Interior, Bureau of Mines Contract SO 122050.

Draper, J.C., 1973. Transportation, *Elements of Practical Coal Mining,* S.M. Casidy (editor), (New York: AIME) pp. 123-154.

European Community for Coal and Steel, personal communication, 1973-74. Washington, D. C.

Federal Register, 1971. May 22: 36 FR 1364; July 15: 36 FR 13143.

_____, 1974. Proposed Amendments 77.215-216. Volume 39, Number 11, Wednesday, January 16, 1974.

5th International Strata Control Conference, 1972. Organized by the British National Coal Board, 32 papers, discussion, and national reports from 11 countries.

General Analytics, Inc., 1974. State of the art of subsidence control, preliminary draft for the Appalachian Regional Commission Grant No. ARC 73-111, EER-119, 100 pp.

Glover, H.G., 1973. The disposal of coal mine spoil in the United Kingdom, *Proceedings of the NATO Advanced Studies Institute on Waste Disposal, and The Renewal and Management of Degraded Environments*, July 13-28, 1973.

Gluckauf, 1972a. Schwebendor Strabbau mit Schreitausbau und Blasversatz auf de Grube Folschviller, v. 108, no. 26, pp. 1238-1243.

_____, 1972b. Zahlen zur Rationalisierung in Steinkohlenbergbau der Bundesrepublik Deutschland, v. 108, no. 15, p. 639.

_____, 1973a. Der Britische Steinkohlenbergbau im Westeuropaischen Vergleich, v. 109, no. 11, pp. 585-593.

_____, 1973b. Der Geschaftsbericht 1972/73 des National Coal Board, v. 109, no. 22, pp. 1110-1116.

_____, 1973c. Der Kohlenbergbau der Bundesrepublik Deutschland im Jahre 1972, v. 109, no. 7, pp. 414-423.

_____, 1973d. Der Kohlenbergbau der Bundesrepublik Deutschland im Ersten Halbjahr 1973, by W. Ortsack, vol. 109, no. 20, pp. 1003-1012.

_____, 1973e. Die Charbonnages de France im Geschaftsjahr 1971, v. 108, no. 24, pp. 1161-1163.

_____, 1973f. Die Charbonnages de France im Geschaftsjahr 1972, v. 109, no. 23, pp. 1175-1179.

_____, 1973g. Die Hohlenimporte der Europaischen Gemeinschaft aus den Landern des Comecon, by H.E. von Scholz, v. 109, no. 9, pp. 494-509.

_____, 1973h. Zahlen aus dem Kohlenbergbau der Bundesrepublik Deutschland fur die Jahre 1971/72 nach Angaben der Statistik der Kohlenwirtschaft e.V., v. 109, no. 4, pp. 284-285.

Gray, R. and H. Salver, 1971. Discussion of Voight and Pariseau's paper, "State of predictive art in subsidence engineering," *Journal of Soil Mechanics and Foundations Division of ASCE*, v. 97, no. SM1, January, 1971.

Gupta, R.N., B. Singh and K.N. Sinha, 1972. Investigations into the compressibility of stowing materials and the parameters affecting it, *Central Mining Research Station, Dhanbad, Research Report No. 55*, 132 pp.

Hambright, J., 1973. Pennsylvania Bureau of Air Quality and Noise.

HR 11500, 1973. Surface Mining Control and Reclamation Bill.

Industrie Minerale Mine, 1970. Aspects statistiques de l'exploitation des houilleres in 1970, August-September, 1972, pp. 57-64.

_____, 1973a. Aspects statistiques de l'exploitation des houilleres en 1971, February, 1973, pp. 25-32.

_____, 1973b. Rapport annual du National Coal Board, February, 1973, v. 55, no. 2, pp. 80-83.

Jacobi, O., 1966. Occurrences, causes, and control of rock bursts in the Ruhr District, *International Journal of Rock Mechanics and Mineral Sciences*, v.3, sec. 3, pp. 205-219.

Joy Manufacturing Company, personal communication, 1973-74, and various articles including: Modern mining methods, by K.E. McElhattan, 8 pp.

Konchesky, J.L. and T.J. George, 1971. Pneumatic transportation of mine run coal underground, *Mining Congress Journal*, December 1971.

_____, 1973. Air and power requirements for the vacuum transport of crushed coal in horizontal pipelines, paper presented to the Joint Materials Handling Conference, Pittsburgh, Pennsylvania, September 19-21, 1973.

Krevelen, D.W., 1961. Coal--typology, chemistry, physics, constitution, (New York: Elsevier Publishing Company) 514 pp.

Lifton, R.J., personal communication, 1973. M.D. and professor of psychiatry, School of Medicine, Yale University, unpublished preliminary report on April 9, 1973 visit to Buffalo Creek.

Luckie, P.T., J.W. Peters and T.S. Spicer, 1966. The evaluation of anthracite refuse as a highway construction material, *Pennsylvania State University Publication No. SR-57*.

Midwest Research Institute, 1973. Methods for identifying and evaluating the nature and extent of nonpoint sources of pollutants, draft report for the Environmental Protection Agency, Contract No. EPA 68-01-1839, Project No. MRI 3774-C.

Mine Safety Appliances Research Corporation, 1973. The uses of foams in underground mines, *Interior Report prepared for the Environmental Protection Agency under Contract No. 68-01-0716*, R.W. Hiltz, project manager.

Mountain Eagle, 1972. Large mine pond gives way at Bethelkorn, October 12, 1972 issue.

National Commission on Materials Policy, 1973. *Material Needs and the Environment Today and Tomorrow*, (Washington, D. C.: U.S. Government Printing Office).

Nicholson, D.C. and W.R. Wayment, 1964. Properties of hydraulic backfills and preliminary vibratory compaction tests, *U.S. Bureau of Mines Report of Investigations 6477*, 31 pp.

_____, 1967. Vibratory compaction of mine hydraulic backfill, *U.S. Bureau of Mines Report of Investigations 6922*, 52 pp.

Palowitch, E.R. and T.J. Brisky, 1973. Designing the Hendrix no. 22 short-wall, *Mining Congress Journal*, v. 59, no. 6, pp. 16-22.

Patterson, R.M., personal communication, 1973-74. Government relations manager, the Dow Chemical Corporation.

Peluso, R.G., 1974. A Federal view of the coal waste disposal problem, *Mining Congress Journal*, January, 1974, pp. 14-17.

Pennsylvania Code, 1962. Pennsylvania Clean Air Act.

_____, 1965. Pennsylvania Clean Stream Law.

_____, 1966. Pennsylvania Mine Subsidence Act.

_____, 1968. Pennsylvania Coal Refuse Disposal Control Act.

Pennsylvania Department of Environmental Resources, 1973. Unpublished survey of active coal mine waste piles in Pennsylvania by staff of the Department during the spring and summer of 1973.

Pennsylvania State University, unpublished. Work of the Department of Mineral Preparation under the sponsorship of the U.S. Public Health Service.

Public Law, 1969. 91-173--Federal Coal Mine Health and Safety Act of 1969.

_____, 1970. 91-604--Clean Air Amendments of 1970.

_____, 1970. 91-611--Flood Control Act of 1970.

_____, 1972. 92-367--National program for dam inspection.

_____, 1972. 92-500--Federal Water Pollution Control Act Amendments of 1972.

Radmark Engineering, division of Rader Pneumatics, Ltd., personal communication, 1973-74. North Vancouver, British Columbia, Canada.

Report of the Tribunal Appointed to Inquire into the Disaster at Aberfan on October 21st 1966, 1968. (London: Her Majesty's Stationery Office) 151 pp.

Sinclair, J., 1963. *Ground movement and control at colliers*, (London: Sir Isaac Pitman & Sons, Ltd.).

Singh, M.M. and W.J. Courtney, 1974. Application of pneumatic stowing in United States coal mines, paper to be presented at the American Institute of Mining Engineers Annual Meeting, February 24-28, 1974.

Starfield, A.M. and S.L. Crouch, 1973. Elastic analysis of single seam extraction, In: H.R. Hardy, Jr. and R. Stefanko (editors), *New Horizons in Rock Mechanics*, (New York: American Society of Civil Engineers) pp. 421-439.

Steele, D.J., 1973. Some impressions of the coal industry of Poland, 1972, *The Mining Engineer*, v. 132, pt. 7, no. 151, pp. 327-339.

Thomson, G. McK. and S. Rodin, 1972. *Colliery Spoil Tips--After Aberfan*, (London: The Institution of Civil Engineers) 60 pp.

Tweedy, D.H., 1973. Recent developments in pneumatic conveying, paper presented to the Northwest Metals and Minerals Conference of the American Institute of Mining Engineers, April 14, 1973.

UNESCO and the Société de l'Industrie Minerale, Paris, 1963. *Atlas of Mining Methods*, group of experts headed by Bohuslav Stocas, (Czechoslovakia: Bohuslav Stocas) 2v., (volume 1, 1964; volume 2, 1966).

United Mine Workers of America, 1974. Preliminary report concerning the multiple fatality at Island Creek's Guyan No. 5 Mine, January 30, 1974, by R. Cooper, Safety Division.

U.S. Comptroller General, 1973. Problems caused by coal mining near Federal reservoir projects, Corps of Engineers (Civil Functions), Department of the Army, Report to the Conservation and Natural Resources Subcommittee, Committee on Government Relations, U.S. House of Representatives, 53 pp.

U.S. Department of Agriculture, 1974. Research on reclamation of deep-mine refuse piles in Pennsylvania, Forest Service and Coal Research Board, W.H. Davidson (researcher), Contract Nos. 68-25 and 69-45.

U.S. Department of the Interior, unpublished a. Environmental effects of underground mining and mineral processing, draft report, 239 pp.

_____, unpublished b. Mineral industry solid waste and our environment, and mineral waste disposal--magnitude and nature of the problem, Bureau of Mines reports by C.W. Gwinn.

_____, unpublished c. 1972 Summary of waste embankments, Health and Safety District No. 1, Bureau of Mines.

_____, unpublished d. Statistics gathered for this committee by the Bureau of Mines.

_____, 1912. Mining conditions under the city of Scranton, Pennsylvania, *Bureau of Mines Bulletin 25*, by W. Griffith and E.T. Conner and a chapter by N.M. Darton, 89 pp.

_____, 1913a. Hydraulic mine filling, its use in the Pennsylvania anthracite fields, *Bureau of Mines Bulletin 60*, by C. Enzian, 77 pp.

_____, 1913b. Sand available for filling mine workings in the native anthracite basin of Pennsylvania, *Bureau of Mines Bulletin 45*, by N.M. Darton, 33 pp.

_____, 1956. Anthracite mechanical mining investigation, *Bureau of Mines Report of Investigations 5290*, by J.C. Hartley, J.D. Cooner, Sr., and R.J. Brennan, 20 pp.

_____, 1964. Summary of burning coal mine refuse banks, *Bureau of Mines Information Circular 8209*, by R.W. Stahl, 39 pp.

_____, 1966. Ignition and control of burning of coal mine refuse, *Bureau of Mines Report of Investigations RI6758*, by J.W. Myers, *et al.*, 24 pp.

_____, 1968. A dictionary of mining, mineral, and related terms, *Bureau of Mines Special Publication*, P.W. Thrush (editor), (Washington, D.C.: U.S. Government Printing Office) 1269 pp.

_____, 1969a. *Minerals Yearbook*, by the Bureau of Mines Staff, (Washington, D.C.: U.S. Government Printing Office) 4v.

_____, 1969b. Pennsylvania anthracite refuse, *Bureau of Mines Information Circular 8409*, by J.C. MacCartney and R.M. Whaite.

_____, 1970a. Mineral facts and problems, *Bureau of Mines Bulletin 650*, (Washington, D.C.: U.S. Government Printing Office) 1291 pp.

_____, 1970b. Utilization of fly ash for remote filling of mine voids, Second Symposium on fly ash utilization, *U.S. Bureau of Mines Information Circular 8488,* by M.O. Magnuson and T.M. Wilkert.

_____, 1971a. Bureau of Mines research programs on recycling and disposal of mineral-, metal-, and energy-based solid wastes, *Bureau of Mines Information Circular 8529,* by C.B. Kenahan and E.P. Flint, 53 pp.

_____, 1971b. Coal refuse fires, an environmental hazard, *Bureau of Mines Information Circular 8515,* by L.M. McNay, 50 pp.

_____, 1971c. Mine subsidence--extent and cost of control in a selected area, *Bureau of Mines Information Circular 8507,* by W. Cochran, 32 pp.

_____, 1971d. Refuse piles, guidelines, and operating procedures pursuant to United States Code of Federal Regulations 77.214 and 77.215.

_____, 1972a. Emergency investigations of coal mine waste embankments, Bureau of Mines and Geological Survey Task Force to Study Coal Waste Hazards, Interim Report.

_____, 1972b. Final environmental impact statement--demonstration--hydraulic backfilling of mine voids, Scranton, Pennsylvania, Bureau of Mines.

_____, 1972c. List of Coal Waste Banks, Bureau of Mines and Geological Survey Task Force to Study Coal Waste Hazards, 288 pp.

_____, 1972d. Preliminary analysis of the coal refuse dam failure at Saunders, West Virginia, February 26, 1972, Bureau of Mines and Geological Survey Task Force to Study Coal Waste Hazards, 42 pp.

_____, 1972e. United States energy--a summary review, 42 pp.

_____, 1973a. Coal mine health and safety manual for coal waste deposits, Mining Enforcement and Safety Administration, 78 pp.

_____, 1973b. Draft environmental statement, proposed regulations governing the disposal of coal mine waste. 30 CFR, part 77, sections 77.215 and 77.216, Press release by the Mining Enforcement and Safety Admistration, November 23, 1973.

_____, 1973c. Methods and costs of coal refuse disposal and reclamation, *Bureau of Mines Information Circular 8576.*

_____, 1973d. Mineral industry surveys--coal, bituminous and lignite in 1972, preliminary release of information pending publication of the *Bureau of Mines Minerals Yearbook,* 77 pp.

_____, 1973e. United States mineral resources, *Geological Survey Professional Paper 820*, D.A. Brobst and W.P. Pratt (editors), (Washington, D.C.: U.S. Government Printing Office) 722 pp.

_____, 1974. Preliminary draft, Physical property data on coal waste embankment materials, *Bureau of Mines Technical Progress Report*, by R.A. Busch, R.R. Backer and L.A. Atkins.

U.S. Environmental Protection Agency, 1971. Control of mine drainage from coal mine mineral wastes, *Water Pollution Control Research Series*, *14010 DDH 08/71*.

_____, 1973. Information on methods to control pollution from mining methods, draft report, Water Quality and Nonpoint Source Control Division, 402 pp.

U.S. Senate, 1973. A review of energy policy activities of the 92nd Congress, Committee for Interior and Insular Affairs.

Vidal, V., 1961. *Exploitation des Mines*, (Paris: Dunod), v. 1, pp. 515-560.

W.A. Wahler and Associates, 1973. Analysis of coal refuse dam failure Middle Fork Buffalo Creek, Saunders, West Virginia, *U.S. Bureau of Mines Contract No. SO 122084*, 2v.

West Virginia Code, Chapter 20, Article 5D, Dam Contract Act, 1973.

_____, Chapter 20, Article 6C, Coal Refuse Disposal Act, 1972.

_____, Chapter 22, Section 4, Article 1, Powers and Duties of the Director of the Department of Mines.

Wild, H.W., 1971. Goals and challenges for research and development work in coal mining from the point of view of Ruhrhohle AG, Gluckauf, v. 107, April 15, 1971, pp. 289-298.

Williams, I., personal communication, 1973. Bureau of Mines, Scranton, Pennsylvania, former president of Blue Coal Company.

Woodruff, S.D., 1966. *Methods of Working Coal and Metal Mines*, (New York City: Pergamon Press) 3v.

Zwartendyk, J., 1971. Economic aspects of surface subsidence resulting from underground mineral exploitation, Ph.D. Thesis, Pennsylvania State University, 411 pp.

APPENDIX A: GLOSSARY OF TERMS

advance mining: The work of excavating as mining goes forward in an entry and in driving rooms; as distinguished from retreat.

advancing longwall: Mining the coal outward from the shaft pillar and maintaining roadways through the worked-out portion of the mine.

angle of draw: In mine subsidence the "angle of draw" is defined as the angle between a vertical line at the extremity of a mined area and the practical limit of measurable subsidence at the surface. It is an empirical quantity and is usually considered to be between 20 and 40 degrees.

backfill: a) Waste refuse, sand or rock used to support the roof after removal of ore from a stope; b) material excavated from a site and used for filling; c) material used to fill a mine opening.

backfilling: The process of placing backfill in a mine opening.

brattice: A partition in a mine passage to confine the air and force it into the working places. Temporary brattices are often made of cloth.

brattice cloth: Fire-resistant canvas or duck cloth used to erect a brattice. A heavy canvas often covered with some waterproofing material, for temporarily forcing the air into the face of a breast or heading; also used in place of doors on gangways.

bone: a) A hard coal-like substance high in noncombustible mineral matter; often found above or below, or in partings between layers of relatively pure coal; b) in the anthracite-coal trade, a carbonaceous shale containing coal; bony coal; c) a layer of hard, impure coal which sometimes grades uniformly into the adjacent softer coal and sometimes is sharply separated from it.

British Thermal Unit: Heat needed to raise 1 pound of water 1° F (equal to 252 calories). Symbol BTU.

bump: Sudden outbursts of coal and rock that occur when stresses in a coal pillar, left for support in underground workings, cause the pillar to rupture without warning, sending coal and rock flying with explosive force.

bunker: A bin for the storage of materials; the lowermost portion is usually constructed in the form of a hopper.

cave: To allow the roof to fall without any retarding supports or waste packs.

caving: The practice of encouraging the roof over the mined area to collapse freely so that it fills the waste area. In metal mining, caving implies the dropping of the overburden as part of the system of mining.

chock: A square pillar for supporting the roof, constructed of prop timber laid up in alternate cross-layers, in log-cabin style, the center being filled with waste. Commonly called crib.

coal cleaning; coal preparation: These terms refer to the sorting, picking, screening, washing, pneumatic separation, and mixing of coal sizes to the best advantage for (and requirements of) the market. In coal cleaning, only those impurities that are mechanically mixed with the coal are removed by wet or pneumatic (air) cleaning.

crosscut: a) A small passageway driven at right angles to the main entry to connect it with a parallel entry or air course; b) in room and pillar mining, the piercing of the pillars at more or less regular intervals for the purpose of haulage and ventilation. Synonym for breakthrough.

crusher: A machine for crushing rock or other materials. Among the various types of crushers are the ball-mill, gyratory crusher, Hadsel mill, hammer mill, jaw crusher, rod mill, rolls, stamp mill, and tube mill.

dry wall packing: A wall without cementing material. Obsolescent method of supporting underground workings by use of waste rock built into rough walls.

entry: a) In coal mining a haulage road, gangway, or airway to the surface; b) an underground passage used for haulage or ventilation, or as a manway; c) a coal heading: To develop a coal mine in the United States, one or more sets of main entries are driven into the seam. Each set consists of four to eight coal headings, connected at intervals by crosscuts. From these, and usually at right angles, butt entries, three to six in number, are driven at intervals of up to 1,500 yards. Between the sets of butt entries, face entries, three to four in number, are driven at intervals of up to 500 yards to form a block or panel. The entries to split the panels may be 12 to 20 feet wide and at 50 to 100 feet centers. Each entry is made as productive as possible and productivity is often higher in the entry work than in pillar extraction.

face: a) The solid surface of the unbroken portion of the coalbed at the advancing end of the working place; b) a point at which coal is being worked away, in a breast or heading; also working face. The working face, front, or forehead, is the face at the end of the tunnel heading, or at the end of the full-size excavation; c) a working place from which coal or mineral is extracted; d) the exposed surface of coal or other mineral deposit in the working place where mining, winning, or getting is proceeding.

fines: a) Very small material produced in breaking up large lumps of coal; b) in general, the smallest particles of coal in any classification, process, or sample of the run-of-mine material; c) coal refuse fines are those small solid particles including coal, which are smaller than 1/2 mm.

This waste material is a byproduct of the fine coal cleaning circuit of the coal preparation plant.

flushing: Filling mine workings with sand, waste, etc., by hydraulic or pneumatic methods. Generally refers to backfilling inactive mines.

fly ash: Fine solid particles of noncombustible ash with or without combustible particles removed from coal fired boiler chimneys by mechanical or electrostatic separators.

gate: An underground roadway for air, water, or general passage; a gangway.

gob: a) That part of a mine from which the coal has been worked away and the space more or less filled up; b) the refuse or waste left in the mine; c) also called goaf in Britain; d) to store underground, as along one side of a working place, the rock and refuse encountered in mining; e) the waste material so packed or stored underground.

gob pile; gob dump: A pile of mine waste on the surface.

in situ: a) In place; b) in the natural or original position. Applied to a rock, soil, or fossil when occurring in the situation in which it was originally formed or deposited.

lift: a) A certain thickness of coal worked in one operation; b) the extraction of a coal pillar in lifts or slices; c) the distance between the first level and the surface or between any two levels; d) a step or bench in a multiple layer excavation.

longwall: A method of working coal seams believed to have originated in Hampshire, England, towards the end of the seventeenth century. The seam is removed in one operation by means of a long working face or wall, thus the name. The workings advance (or retreat) in a continuous line which may be several hundreds of yards in length. The space from which the coal has been removed (the gob, goaf, or waste) is either allowed to collapse (caving) is completely or partially filled or stowed with stone and debris.

longwall advancing: A system of longwall working in which the faces advance from the shafts towards the boundary or other limit lines. In this method, all the roadways are in worked-out areas.

longwall retreating: a) A system of longwall working in which the developing headings are driven narrow to the boundary or limit line and then the coal seam is extracted by longwall faces retreating toward the shaft. In this method, all the roadways are in the solid coal seam and the waste areas are left behind; b) first driving haulage road and air ways to the boundary of a tract of coal and then mining it in a single face back toward the shaft.

methane: CH_4, carbureted hydrogen or marsh gas or firedamp; formed by the decomposition of organic matter. The most common gas found in coal mines. It is a tasteless, colorless, nonpoisonous, and odorless gas; in mines the presence of impurities may give it a peculiar smell. Its weight relative to air is 0.555 and may therefore form layers along the roof and occupy roof cavities. Methane will not support life or combustion; with air, however, it forms an explosive mixture, the reaction being: $CH_4 + 2(4N_2 + O_2) = CO_2 + 2H_2O + 8N_2$. The gases resulting from a methane explosion are irrespirable. Methane is nontoxic, and its breathing causes ill effects only where the air is so heavily laden with it that oxygen is supplanted.

pack: a) A pillar, constructed from loose stones and dirt, built in the waste area or roadside to support the roof; b) waste rock or timber support for roof of underground workings or used to fill excavations. Also called fill.

panel: a) A large rectangular block or pillar of coal; b) a method of working whereby the workings of a mine are divided into sections, each surrounded by solid strata and coal with only the necessary roads through the coal barrier; c) a system of coal extraction in which the ground is laid off in separate districts or panels, pillars of extra size being left between; d) a group of working places, usually operated as a unit, and separated from others by large pillars of coal.

permissible: a) Means completely assembled and conforming in every respect with the design formally approved for use in gassy and dusty mines; b) in the U.S., machine or explosive is said to be permissible when it has been approved by the U.S. Bureau of Mines for use underground under prescribed conditions. Not all flameproof machinery is permissible but all permissible machinery is flameproof.

pillar: a) An area of coal or ore left to support the overlying strata or hangingwall in a mine. Pillars are sometimes left permanently to support surface works or to isolate old workings containing water. Others are extracted at a later period; b) the part of coal left between the individual rooms and entries in room-and-pillar mining.

pillar robbing: The removal of the coal pillars between rooms or chambers.

productivity: Output (usually measured in tons of coal recovered) per man shift. Productivity will vary with the degree of mechanization and multishift working.

recovery: a) The proportion or percentage of coal or ore mined from the original seam or deposit; b) the percentage of in-place coal removed.

reddog: a) Material of a reddish color resulting from the combustion of shale and other mine waste in dumps on the surface; b) burned coal refuse.

refuse: Waste material in the raw coal which is removed by cleaning. Also called tailings, culm, slate, slay and gob.

retreating system: a) A method of working a mine which is designed to allow a stope to cave soon after it is worked out, thus relieving most of the weight on the supports in adjacent stopes; b) a method of extracting coal or ore by driving a narrow heading to the boundary, then opening out a face and working the deposit backwards towards the shaft, drift, or main entry; c) a stoping system in which supporting pillars of ore are left while the deposit is worked outward from shafts toward a boundary, these pillars being removed (robbed) as the work retreats toward the shaft and the unsupported workings are abandoned and left to cave in.

room: a) Space driven off an entry in which coal is produced. Rooms may vary in width from 14 to 45 feet and in depth from 50 to 300 feet, depending on depth of overburden, underground conditions, and seam thickness; b) a place abutting an entry or airway where coal has been mined and extending from the entry or airway to a face.

room and pillar: A system of mining in which the distinguishing feature is the winning of 50 percent or more of the coal or ore in the first working. The coal or ore is mined in rooms separated by narrow ribs or pillars. The coal or ore in the pillars is won by subsequent working, which may be likened to top slicing, in which the roof is caved in successive blocks. The first working in rooms is an advancing, and the winning of the rib (pillar) a retreating method. The rooms are driven parallel with one another, and the room faces may be extended parallel, at right angles, or at an angle to the dip.

run-of-mine rock: This is partially crushed, usually dry rock discharged directly from mine workings produced primarily in the sinking of the shaft or slope; bottom rock grading along haulage ways; and top rock removal for clearance.

short ton: A unit of weight that equals 20 short hundredweights or 2,000 avoirdupois pounds. Used chiefly in the United States and Canada.

shortwall: a) A method of mining in which comparatively small areas are worked separately, as opposed to longwall; b) a length of coal face intermediate between that of room and pillar faces and a normal longwall face.

slate: A coal miner's term for any shale or slate accompanying coal; sometimes applied to bony coal.

slime; slimes: A material of extremely fine particle size encountered in the processing of coal.

slue: To turn, twist, or swing about. To slide and turn or slip out of course. It is necessary at intervals to stop machines and straighten them, or "slue" them, as called by miners.

slurry: a) A liquid mixture of crushed or other finely divided material and water; b) the fine carbonaceous discharge from a colliery washery. All washeries produce some slurry which must be treated to separate the solids from the water in order to have a clear effluent for reuse or discharge. Also, in some cases, it is economical to extract the fine coal from the effluent.

stacker: A conveyor adapted to piling or stacking bulk materials, including coal or coal waste.

stowing: A method of mining in which the coal of the seam is removed and waste is packed into the space left by the mining.

strain: a) Deformation resulting from applied force; within elastic limits strain is proportional to stress; b) the change in length per unit of length in a given direction.

stratum (pl. strata): A layer, bed, or member of a series of layered materials or rock.

stress: a) Force per unit area, often thought of as force acting through a small area within a plane; b) the force that results in strain.

strip packing: An arrangement of alternate packs and wastes built in a direction parallel to the gate roads in longwall conveyor mining.

subsidence: a) The lowering of the strata, including the surface, due to underground excavations; b) surface caving or distortion due to effects of collapse of mine workings; c) a sinking down of a part of the earth's crust.

subsidence area: The area affected by subsidence over areas where minerals or other substances have been removed by mining. The area usually is larger than the mined out area below.

sump: a) Any excavation in a mine for collecting or storing water; b) an excavation made underground to collect water, from which water is pumped to the surface or to another sump nearer the surface.

tailings: a) The parts, or a part, of any incoherent or fluid material separated as refuse, or separately treated as inferior in quality or value; leavings; remainders; dregs; b) those portions of washed ore or coal that are regarded as too poor to be treated further.

tailings dam: An impoundment for tailings disposal.

tipple: Originally the place where the mine cars were tipped and emptied of
their coal, and still used in that sense, although now more generally
applied to the surface structures of a mine, including the preparation
plant and loading tracks.

(coal) waste pile: Any deposit of coal refuse placed on the surface and intended for temporary or permanent disposal. Synonyms include refuse pile,
refuse bank, culm bank, gob pile, slate bank, and coal wastebank.

The foregoing definitions have been adopted with modifications from the U.S.
Department of the Interior, 1968, Dictionary of mining, mineral and related
terms, Bureau of Mines Special Publication, P.W. Trush/editor, (Washington,
D.C.: U.S. Government Printing Office) 126 pp.

APPENDIX B: EFFECTS OF UNDERGROUND COAL MINING ON SURFACE SUBSIDENCE AND ON UNDERGROUND ROOF CONTROL

STRATA DISTURBANCE DUE TO MINING

The effects of mining of a coal seam on the neighboring rock mass may be illustrated by considering the very simple situation in which a horizontal coal seam of thickness T and of large area l extent is located at a depth (H) below the level surface (Figure B.1). The weight of the overlying rock exerts a vertical pressure p on the coal seam. As a rule of thumb, p is taken to be 0.25 kg per square centimeter per meter depth (1.15 lb/in.^{2}/ft)*, (or 1.15H lb/in.^{2} at depth H), so that at 200 m (625 ft) depth, the pressure is roughly 50 kg/cm^2 (720 lb/in.^{2}). If, in mining, 50 percent of the coal is removed (as e.g., rooms), then the total weight of overlying rock must be carried by the 50 percent remaining (in the pillars). Thus, the pressure will be doubled so that, on the average each pillar must be able to withstand 100 kg/cm^2 (1450 lb/in.^{2}) without collapse. If the pillars are adequate then the rooms and mine roadways will remain stable and there will be no perceptible effects on the overlying rock or the surface.

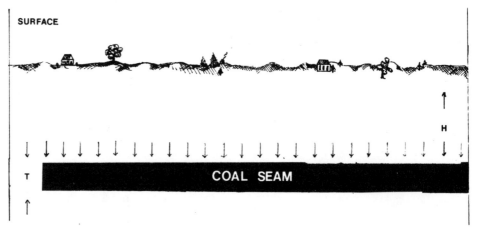

FIGURE B.1   Uniform Pressure On Coal Seam Before Mining.

---

*A cube of rock 1 ft on a side will weigh approximately 150-180 lb. Assuming 170 lb. as an average, then the verticle pressure across the bottom face exerted by the weight of the cube will be 170 lb. over the 1 ft^2, or 170 lb. over 144 in.^{2}. This corresponds to 170/144 = 1.15 lb/in.^{2}.

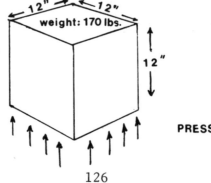

126

If the amount mined is such that the pillars crush or, as in longwall mining, the coal is completely removed over large portions of the mine then the roof and overlying rock will collapse or in some cases may settle gradually into the mined out space. The extent to which the rock movements underground are transmitted to the surface depends principally on such factors as the geology of the region, and the strength, thickness and lamination of the rock strata above the mine. Concern over damage resulting from subsidence due to undermining in builtup areas has led to considerable research and field measurements in Europe over the past 40-50 years, from which some useful general empirical concepts and rules have been derived. More recently, fundamental studies have revealed that there may be severe limitations in the general applicability of the empirical concepts, but they do allow a reasonably simple explanation of the subsidence effects that can occur.

The rock overlying coal seams occurs in layers and tends to deflect under its own weight as the coal is removed or pillars crush due to excessive pressure.

Consider the situation shown in Figure B.2. The coal has been extracted over the space AB such that there is little or no resistance to the deflection (or collapse) of the mine roof into the opening. This in turn allows deflection of the overlying rock strata, the movement being transmitted eventually to the surface, to form a more or less basin-shaped depression--or "subsidence trough," CED. The original flat surface CD is "stretched," (or "strained"), horizontally to form the basin CED. Any construction in intimate contact with the ground surface (e.g., houses, roads, rails, sewers, utility lines) or features such as lakes and canals will be subjected to these ground movements and may be seriously damaged.

Obviously, the total volume of the surface basin cannot exceed the total volume of the underground excavation. Should the rock be so weak as to approach the consistency of a fluid e.g., weak clays, then it could flow so as

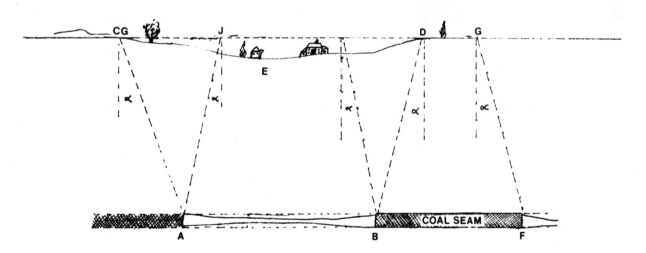

FIGURE B.2 Subsidence trough and angle of draw concept in determining extent of effects of mine subsidence.

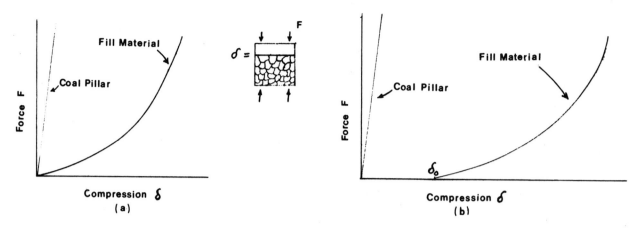

FIGURE B.3 Compression behavior of fill (a) with no initial gap (b) with initial gap δ_o between roof and top of fill.

to fill the mined out areas. In this case the surface depression would essentially equal the volume of the excavated coal. In actual cases, as the rock subsides into the voids it tends to fill the space with a somewhat loose assemblage, with gaps of broken rock slabs, so that the excavated space is not completely filled. The volume of the subsidence basin is then somewhat less than the volume of the excavated coal. If the excavated space is backfilled in some manner then the subsidence trough volume and associated ground movement will be reduced still further, depending on the amount and inplace density of the backfill material.

The maximum depth of the subsidence trough can reach an upper limit equal to the thickness (T) of the excavated coal seam, but will tend to be somewhat less, depending on the amount of backfill, remnant coal pillars, etc. Tests on backfill material or, indeed, on any loosely consolidated material, show that initially it offers little resistance. (There are small voids between the backfilled particles which result in compression after placement.) A typical compression curve is shown in Figure B.3. Thus, backfill material, even when tightly packed does not prevent the development of subsidence but rather limits its eventual magnitude.

Zwartendyk* (1971) quotes the following values for the maximum vertical subsidence of the surface S_m observed for various coal fields in Europe where longwall mining was practiced and the roof was allowed to collapse or "cave" without backfilling.

---

*Zwartendyk's thesis contains an excellent nontechnical review of the pre-1970 "state of the art" of subsidence theories.

TABLE B.1 Maximum Vertical Subsidence Observed in Europe

	S_m/T	
Great Britain	0.90	
Holland	0.91-0.93	
Germany	0.95	T is the seam
France	0.85-0.90	thickness.

He also cites a suggestion that the corresponding ratio for U.S. longwall mining might be less than 0.75, because of the practice of leaving the entry pillars unmined. Dahl (1972) reports field measurements that tend to support this lower value. U.S. experience with pillar extraction in room-and-pillar mining indicates an S_m/T ratio of 0.4 to .75 (Gray and Salver, 1971).

Backfilling of the space by various methods can substantially reduce the surface displacement. In Europe (Zwartendyk, 1971) it has been found, for example, that S_m/T is reduced as follows:

Pneumatic or mechanical stowing S_m/T = 0.45-0.50
Hydraulic stowing S_m/T = 0.25-0.35

DEVELOPMENT OF SUBSIDENCE WITH TIME

The length of time between underground extraction and the appearance of subsidence at the surface depends on the depth of mining, and the nature of the rock strata above the mined areas. In most cases, for mines less than 1,000 feet deep the initial effects usually appear within 1-2 days and develop rapidly, being essentially complete within several weeks. There are, however, examples of subsidence occurring gradually over a much longer period, of the order of 1-2 years. In other situations, such as abandoned mines where the roof is supported by remnant pillars, it is possible for the pillars and roof and floor rock to weaken over many years e.g., by water or atmospheric attack, and eventually they collapse causing surface subsidence.* In such cases, the use of backfill may be effective in preventing pillar degradation and collapse.

ANGLE OF DRAW

European surveyors have made extensive measurements over the past 40 or more years correlating the underground extraction pattern with the associated

---

* The final collapse may be triggered by some abnormal event, as in the case of the floods in Scranton associated with Hurricane Agnes in 1974.

surface effects of mining. This has led to the so-called "angle of draw" empirical subsidence rule. The rule has been shown to be suspect for deep mines (1500 ft-500 m) or deeper, but appears to describe adequately effects observed in shallow mines. The angle of draw is determined for a particular area by observing the extent, e.g., CED, (Figure B.2) of *discernible** surface strain or vertical subsidence. It is assumed that as the face B moves forward, then the trough limit D will advance equally. Thus the line BD is assumed to be characteristic of the strata and the angle, and (Figure B.2) is termed the "angle of draw." In European coal fields, it has been found that the angle of draw is about 35° to 40°.

Using the angle of draw it is simple to define the area of coal pillar support needed to protect a portion of the ground surface, e.g., one encompassing some vital facility that should not be disturbed by subsidence. The surface region to be protected (e.g., DG) is defined and projected to the seam level at the angle of draw. Thus the pillar BF (similarly calculated in each direction) is needed to protect the surface DG. Projection of the angle of draw over the excavated area (lines BH and AJ) is assumed to define the surface region (JH) liable to maximum subsidence, S_m. If the two internal lines BH and AJ intersect below the surface, this implies that the surface will undergo less than maximum subsidence. Filling of the excavated area will obviously reduce somewhat the volume of voids into which the overlying rock can move. The extent to which the rock movement and, in turn, subsidence, will be reduced depends on the compressibility of the backfill, i.e., the rapidity of increase of the force resisting backfill compaction (Figure B.3). If the excavated area is only partially filled, then the resistance will be small, so that a large amount of rock movement and subsidence can take place. The density with which fill can be stowed in place varies and the amount of subsidence varies similarly.

The need to limit surface subsidence depends on several factors, all of which can be considered overall as the economic and social cost of allowing the surface to undergo severe subsidence. In Europe, for example, even in built up areas, the extracted region usually is not backfilled but is allowed to cave because it is considered less costly to pay compensation for damage and undertake surface treatment than to use backfilling to reduce subsidence. In some cases, facilities such as roads, bridges, houses and other structures have been specifically and successfully designed to withstand anticipated subsidence from undermining. In the United States some states have laws related to subsidence, one of which (Pennsylvania's Land Conservation Act of 1966) is based on empirical subsidence formulae (Cortis, 1969).

UNDERGROUND ROOF CONTROL

Surface subsidence is one manifestation of the fact that underground mining results in ground movements. It is the end result of deformations originating in the mine roof immediately overlying the coal seam that is being

---

*This limit increases (i.e., the basin widens) with increased precision of surveying equipment.

extracted. Roof-falls are a major cause of deaths and injury to miners, so that effective roof control is a subject of major concern in coal mining.

Extraction of coal to form a room in room-and-pillar mining causes the roof to deflect downwards (to "sag") and the floor to deflect upwards (to "heave"). The rock above and below the coal is usually layered and deflects more or less as a beam. The deflection depends on the width of opening, thickness and stiffness of each layer, and the degree of frictional inter-locking between adjacent roof layers. Localized tensile stresses are induced by the deflection. Since rock is much weaker in tension than in compression, it is possible that even relatively low tensile stresses may cause the roof to fracture and perhaps collapse. Separation between layers may also result due to the weight of each layer. If not properly supported, one or more layers of the roof may fall into the mine void and in some cases may result in an arch of up to 60° or so above the opening, as shown in Figure B.4.

Prior to the introduction of mechanized mining it was customary to erect wooden sets or props to hold up the mine roof. However, these tend to ob-struct free movement of equipment in the mine rooms and it is now standard practice in most mines to use roof-bolts or "rockbolts" to reinforce the roof rock-layers so that they safely span the room (see Figure B.5). A rockbolt is a steel rod 1/2-1 inch in diameter and 3-6 feet or in some cases even 10 or 12 feet long that is inserted into a hole drilled into the mine roof and held in place either by a mechanical anchorage at the bottom of the hole or by resin grouting along the length of the bolt. Mechanically anchored rockbolts are tensioned by tightening a nut against a bearing plate on the mine roof, while resin grouted bolts generally are untensioned. Mechanically anchored bolts depend upon maintenance of their tension to be effective (presumably by keep-ing the roof rock-layers squeezed together), while grouted bolts act to preserve the integrity of the layers by preventing them both from sliding past one another and from separation.

Rockbolts are installed during advance mining as part of the mining cycle, except during pillar recovery work when portions of the roof may be allowed to cave. When complete caving is practiced, it is desirable to get a "clean break" of the roof to relieve some of the load on the pillars. Clean roof breaks also tend to make pillar loading more uniform, which reduces bend-ing stresses in the roof and leads to generally improved mine roof conditions.

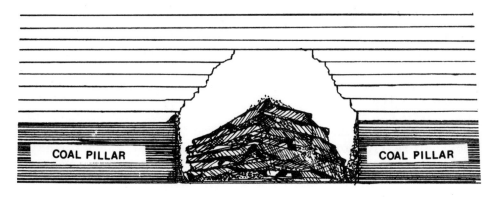

FIGURE B.4 Collapse of coal seam roof.

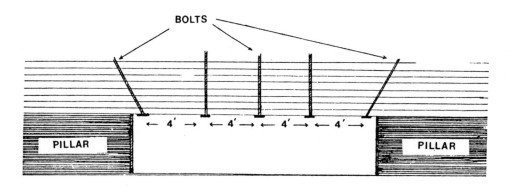

FIGURE B.5 A typical pattern of coal mine roofbolting.

Although roof-falls between coal-pillars are of serious concern and have caused fatal accidents they are nevertheless of relatively limited extent compared to the more massive ground movements produced during pillar extraction and longwall mining operations.

If, as intended in partial extraction systems, the pillars are large enough so as not to crush, then local roof-falls will cause no perceptible surface subsidence. The massive roof collapse ("caving") associated with pillar extraction or longwall mining can result in major subsidence. Roof and ground-control problems associated with these mining operations may be illustrated by reference to Figure B.6. Figure B.6(A) represents a cross-section showing an unmined coal seam under a uniform cover of rock, so that vertical pressure is uniform and of intensity equal to p. Figure B.6(b) represents a similar cross-section of the same coal seam after extensive longwall mining has taken place and the roof behind the supported surface has been allowed to collapse into the excavated space. Removal of the coal causes the roof above to partially fracture, but collapse is prevented by the face supports. Behind the rear support, the roof breaks up into slabs and pieces of various sizes, rotating and sliding into the void. The volume occupied by the broken rock is substantially greater than that before it collapsed due to the voids in the caved rock. Thus, the rock breakup occurs until either a competent rock layer is reached that can span the unsupported area, i.e., from the face to some distance into the gob at which significant support is developed, say at (D) or until the height is just sufficient to accommodate the broken rock mass.*

As the face (and cave) advances, the strata overlying the cave deflect and progressively compact the caved rock. As with the backfill material described in Figure B.3 the load carrying capability of the caved rock increases

---

*Thus, if the volume occupied by the broken rock is 40 percent greater than the same rock before fracture, then, if we assume that the void height is T and that the rock moves only vertically, then the height of cave T_c will be given by $\dfrac{T}{T_c} = 0.40$ or $T_c = 2.5\ T$.

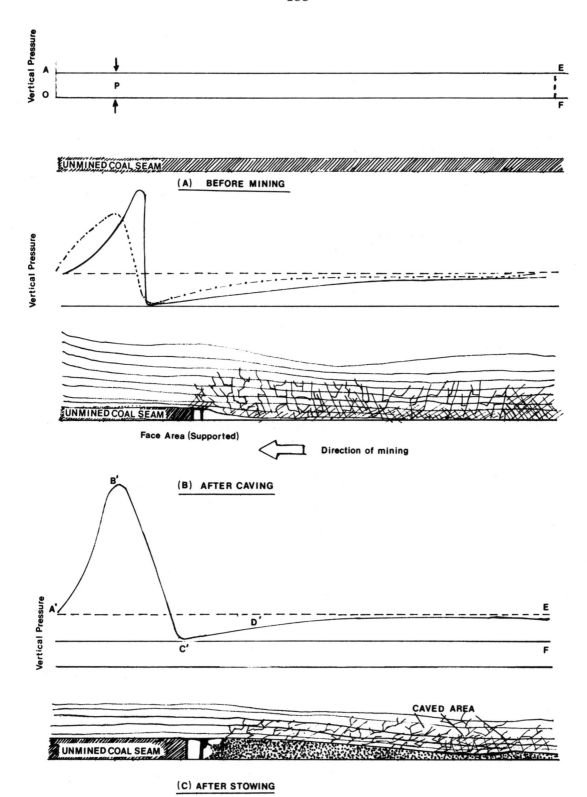

Vertical Pressure

A   P   E

O   F

UNMINED COAL SEAM

(A)   BEFORE MINING

Vertical Pressure

UNMINED COAL SEAM

Face Area (Supported)   Direction of mining

B'

(B) AFTER CAVING

Vertical Pressure

A'   D'   E

C'   F

CAVED AREA

UNMINED COAL SEAM

(C) AFTER STOWING

FIGURE B.6   Pressure Redistribution Produced by (B) Caving &
(C) Stowing.

with compaction so that eventually, when the face has advanced sufficiently, the caved rock supports a pressure almost equivalent to the premining value, p, at which value the caved rock is able to support the overlying rock with no further downward movement.

The variation of pressure distribution from ahead of the face into the gob is shown by the full line ABCDE in Figure B.6(B) with the previous uniform pressure (p) line shown for reference. Extraction causes the overlying strata to deflect and exert extra pressure (peaking at B) on the coal just ahead of the face. This pressure peak is usually referred to as the "front abutment" pressure. Immediately behind the face in the working area the pressure drops to a low value corresponding to that offered by the resistance of the face supports. Moving progressively further into the caved area the pressure increases with compaction of the caved rock, as described above. Since the total weight of rock overlying the coal seam is unchanged, then the total force (i.e., pressure x plan area of the seam) is unchanged by mining. Thus, the area represented by the mean pressure p and the seam length AE..... viz area OAEF in Figure B.6(A) must equal the area under the curve OABCDEF in Figure B.6(B). In other words, the increase in force supported by the face (represented by the region peaking at B) must equal the decrease in force carried by the caved rock (represented by the trough region bottoming at C). Thus, more rapid buildup of the load carried by the caved material (i.e., reduced compressibility of the caved rock or filled material) will reduce the magnitude of the pressure rise ahead of the face.

Figure B.6(C) represents the situation in which the caved area is back-filled with a compressible fill having characteristics such as shown in Figure B.3. The overall effect of the backfill on the roof behavior is to limit the extent to which caving can occur, and to reduce the compressibility, resulting in a somewhat greater portion of the total load being carried by the caved area and a more gradual deflection of the overlying rock strata. This in turn tends to reduce the peak pressure (B') ahead of the face, but may widen the zone of high pressure. The hypothetical* pressure profile for the backfilled situation AB'C'D'E' is shown as a solid line in Figure B.6(C) and also as a dotted line in Figure B.6(B), for ease of comparison with the profile ABCDE. Fracture of the mine roof in the face area is not eliminated by backfilling, but it is generally considered that backfilling improves face roof conditions compared to those obtained in caving. However, development of the intensive support systems used with mechanized mining tends to improve roof conditions in caving situations so that this advantage of backfilling is perhaps not as significant as with nonmechanized longwall caving operations. Also, the pressure in the coal face is considered an important aid in breaking up the coal as it is cut or plowed for loading onto the face conveyor.

Jacobi (1966) indicates that backfilling may tend to *increase* rather than decrease (as suggested above) the pressure peak at the coal face compared to that developed with complete extraction (including all entry pillars) and

---

*No actual pressure measurements have been made to *compare* backfilling and caving distributions, but measurements made by Jacobi (1952) at the Neümuhl mine in Germany, support the general shape of the pressure distributions shown in the filled caved area behind a face.

caving. He considers that the compressibility of pneumatically stowed material is higher than that of caved rock, stating that at a depth of 800 m a pneumatically stowed gob with a minimum dimension of 500 m carried only 38 percent of the overburden load, with the remainder "arched over" to the mine face. Apparently, then, the presence of highly compressible backfill hinders the caving process, which normally acts to decrease the stress concentrations at the mine face. In other words, caved rock supports *more* of the overburden load than pneumatic backfill.*

It would also appear that the situation described by Jacobi will change with time. Thus, if the caved rock is less compressible then the total ground movement over the caved area, and in turn the total surface subsidence, should be reduced. However, it is universally accepted that caving results in greater subsidence than any backfilling system. It could well be that the initial compressibility of caved rock is less than that of backfill thus giving the reduced pressure peak claimed by Jacobi--but that the compressibility increases with time (e.g., due to water action in the voids) so that the *final* subsidence is greater than with backfilling. From the point of view of underground mining, the reduced initial pressure peak would likely improve roof support conditions, i.e., caving would be preferable to pneumatic stowing in rapidly advancing faces.

---

*A slightly different interpretation of this could be given as follows: Caving allows some of the overburden rock to detach itself thus reducing the mass contributing to the pressure peak at the mine face. The detached (i.e., caved) rock then acts to support the remaining overburden, further reducing the pressure peak. There is no doubt that the mine face abutment pressure appears to be significantly higher before caving (i.e., when hangup occurs) than after.

APPENDIX C: BASIS OF COMMITTEE'S COST ESTIMATES

TABLE A Mine Stabilization Projects Utilizing Coal Mine Refuse
Northern Anthracite Field, Pennsylvania

Year Completed	Location (Scranton unless noted otherwise)	Method of Flushing Backfill	Volume Filled (yd^3)	Total Cost	Cost Per Cubic Yard
1956	Morgan Slope mine-Pittson	Blind	271,122	$ 239,974	$0.89
1958	Strip Backfill--Foote Ave.-Pittson	Strip backfill	70,348	146,884	2.09
1960	Downtown Pittson	Blind	3,450,000	79,500	--
1961	Oxford Shaft Mine	Controlled	50,008	82,372	1.65
1961	Oxford Shaft Mine	Controlled	23,985	35,187	1.47
1961	Downtown	Controlled	49,613	74,780	1.51
1962	Monroe & Wolston Ave.	Controlled	169,392	240,165	1.42
1962	N.W. of Oxford Shaft Mine	Controlled	92,680	130,531	1.41
1962	S.W. part of town	Controlled	54,931	75,274	1.37
1963	Meridian & 8th	Blind	61,977	85,975	1.39
1963	Downtown	Controlled	399,368	445,431	1.12

TABLE A (Continued)

1964	Winfield & Prospect Ave.	Con-trolled	10,181	$ 23,753	$ 2.33
1964	Parrish Slope	Blind	4,738	75,386	15.91
1965	GAR	Con-trolled	246,478	585,599	2.38
1965	Tripp Slope	Con-trolled	291,077	534,534	1.84
1965	Eynon Street	Blind	119,836	296,696	2.48
1967	Plymouth	Blind	750	28,000	--
1968	Pine Brook	Con-trolled & blind	449,912	858,865	1.72
1969	Morse School	Con-trolled & blind	269,306	946,474	3.65
1969	Coaldale	Blind	62,759	154,387	2.46
1969	Spring Street	Con-trolled & blind	98,718	154,387	6.76
1969	Heights West	Con-trolled & blind	173,477	1,033,469	5.96
1970	Central City	Con-trolled & blind	632,374	1,306,647	2.07
1971	West Central City	Con-trolled & blind	353,818	1,480,054	4.16
1972	Green Ridge Part 1	Blind	300,000	1,739,670	5.80

TABLE A (Continued)

| 1972 | Green Ridge Enlarged | Blind | 451,117 | $2,165,915 | $4.80 |
| 1973 | Green Ridge Part 2 | Blind | 500,000 | 2,157,245 | 4.31 |

Source: Data from Appalachian Regional Commission, U.S. Bureau of Mines.

TABLE B Labor Cost per Man-Shift

The committee's estimate of labor cost per man-shift of $66 is based upon the following breakdown:

$ 58.00	wages and fringes
3.50	payroll taxes
3.50	workman's compensation 20¢/ton
.26	initial cost (assume 20% turnover, $200 per person)
.50	administrative cost
$ 65.76	

The committee did not include the following costs in its estimate of labor costs:

$ 13.50	United Mine Workers Royalty on production (80¢/ton)
2.00	Federal Black Lung benefits (3.3% of payroll dollars)
5.00	State Black Lung benefits (10% of payroll dollars)
$ 20.50	

NOTE: The United Mine Workers Welfare Fund royalty has not been added to the committee's estimate of labor cost per man-shift of stowing because it is based on the amount of coal extracted and not on the amount of material stowed. It can be assumed that a company will be assessed for the costs of hospitalization and retirement for the nonproductive workers in a mine by an increase in the royalty.

The black lung benefits contained within the Coal Mine Health and Safety Act of 1969 which were originally carried by the Federal and State governments will be carried by the coal operator. Coal companies are already paying for this fund.

TABLE C   Capital Recovery Factor (CRF)

In order to calculate the cost per year to a coal operator of investing capital in equipment for underground disposal, the following assumptions were made:

Estimated useful life of asset	10 years
Depreciation	10 years
Depletion allowance	10%
Capital charge to allow a discounted cash flow rate of return of:	15%
The capital charge includes an income tax rate of:	50%
(Federal 48%; State 2%)	
Investment tax credit	none
Capital Recovery Factor	25.88%

At a capital charge of 10% and an investment tax credit of 7%, the Capital Recovery Factor would be 21.15%.

TABLE D   Committee Estimate of Costs for Surface Disposal   (Breakdown of Table 4.4 from Chapter 4)

The cost estimates of Table 4.4 are based on the following costs rounded to three significant figures:

Size of the operation:

```
    3,000 tons of refuse disposed of per day (1,000 tons per shift)
  690,000 tons of refuse disposed of per year
   69,000 in a slurry impoundment
  621,000 on a waste pile
2,070,000 tons of clean coal produced per year
```

(1)   Land Acquisition

A hollow one mile in length and 3/4 mile in width with a gradual rise of 400-600 feet would cover approximately 500 acres.   In the hilly areas of West Virginia this land would sell for $150-200/acre whereas flat land would cost $1,000-1,500.   In more populated areas the hollows might be more expensive than the $150-200/acre cited for rural West Virginia.   The committee estimates that a preparation plant handling 2,070,000 of clean coal and 690,000 tons of refuse per year would be able to buy a 500 acre hollow to handle 20 years of accumulation of waste for $500/acre or less. $500 x 500 acres = $250,000.   Because it is assumed that the value of the land will be equal or greater than the price which was paid for it, only the 15 percent interest is charged as a capital expense.

$250,000 x 0.15 = $7,500/year

(2)   Surface Disposal System

Capital

Twin-Engine Scraper Cat 657B	$210,000
1500' of Conveyor Belt @ $200/ft	300,000
150-Ton Rock Bin	95,000
Foundations	3,200
Miscellaneous Electrical	5,700
D8 Dozer w/Roller	95,000
TOTAL capital expenditure	$708,900

Capital expenditure per year $708,900 x 0.2588 CRF = $183,463/year

Labor

3 men/day operation of scraper
1/2 man/day maintenance
1/2 man/day roller
   TOTAL: 4 men/day at $66 x 230 days/year = $60,720

TABLE D (Continued)

Operation, Supplies and Maintenance
D8 dozer -- $20/hour operates 4 hours/day
         4 hours x $20 x 230 days/year = $10,120/year
657B scraper -- $35/hour operates 21 hours/day
         21 hours x $35 x 230 days/year = $169,050
Belt operation -- $.05/ton mile for 1500 feet
         1500/5280 x .05/ton mile = $.014/ton
         .014 x 621,000 tons/year = $8,694/year
Ditching of perimeter of the disposal area (7 miles) every 3 years =
         $11,100
TOTAL operating costs = $207,244

(3) Reclamation

Contract Charge

Topsoiling -- removal and replacement $1.35/cubic yard of topsoil
  6,200 cubic yards/acre x 4 acres/year = 24,800 cubic yards
$$\frac{24,800 \text{ cubic yards} \times \$1.35}{690,000 \text{ tons of refuse}} = \$.0485$$
Revegetation -- from EPA and Office of Coal Research
  $375 per acre x 4 acres/year = $1,500
$$\frac{\$1,500 \text{ per year}}{690,000 \text{ tons}} = \$.002/\text{ton refuse}$$

(4) Disposal of Fines

Capital
$500,000 to build an environmentally acceptable and safe impoundment
conforming to the proposed Federal regulations for impoundments that
has the capacity of handling 69,000 tons of fines per year.
$500,000 x .2588 CRF = $129,400/year

Labor
50 man days/year x $66 = $3,300

Maintenance
Hydraulic transport for 1 mile of 69,000 tons of fines/year @0.05/ton
mile. 0.05/ton mile x 1 mile x 69,000 tons of fines = $3,450/year

(5) Loss of Mine Productivity
There is no loss of productivity to the mining operation associated
with surface disposal.

TABLE E   Committee Estimate of the Cost of Pneumatic Backfilling (breakdown and explanation of Table 4.5a, b, and c from Chapter 4)

The cost estimates of Table 4.5 are based on the following costs rounded to three significant figures. The three examples (a, b, and c) are estimates of different quantities of material to be stowed. In all three examples the following conditions have been assumed:

The longwall panel of the committee's model is 450 ft x 3000 ft x 5 ft. The volume of this panel that would be filled for refuse disposal would not include the entries to the panel. Thus, although the gross volume of the longwall panel is 450 x 3000 x 5, the volume to be filled is 438 ft x 2960 ft x 5 ft = 6,482,400 ft^3. It is assumed that a ton of refuse fills a void of one cubic yard; hence each panel will accommodate 259,296 tons of waste material.

At 3000 tons per day raw longwall production, a panel will be mined out in 86.43 days.

$$\frac{259,296 \text{ raw tons/panel}}{3,000 \text{ raw tons/day}} = 86.43 \text{ days/panel}$$

20 days are required to move and set up the longwall in a new panel, therefore, a total of 106.43 days are devoted to each panel. Thus, in a working year of 230 work days, 2.16 panels will be mined. Of these 230 work days, 186.8 are production days (2.16 x 86.43) and 43.2 are moving days (2.61 x 20). [Figure C.1 is an approximation of production fluctuations during two 3-month periods.]

The quantity of refuse required to fill 2.16 panels is 560,079 tons (2.16 panels per year x 259,296 tons per panel) of which the refuse from longwall itself accounts for 140,028 tons per year (2.16 x 64,824). Thus, 420,051 tons per year of refuse from rest of mine can be accommodated in the longwall panel (560,079 tons capacity minus 140,028 longwall refuse).

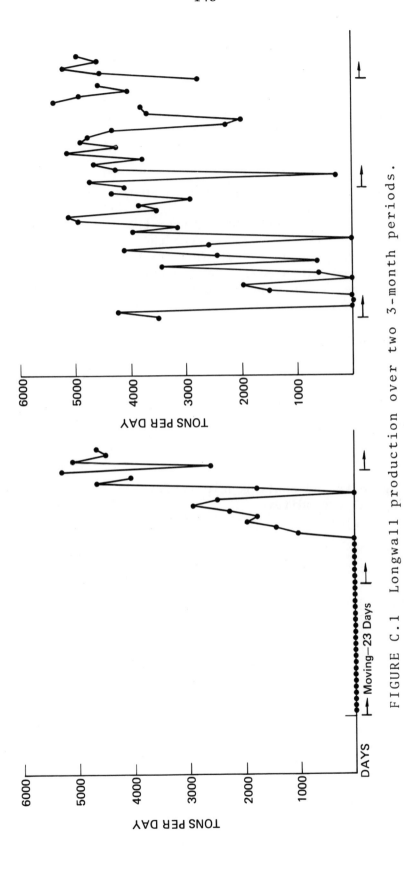

FIGURE C.1 Longwall production over two 3-month periods.

TABLE E(a)   Pneumatic Backfilling (breakdown of Table 4.5a)

Estimate of the cost of pneumatic backfilling behind a longwall face in an equilibrium situation where the waste produced by the entire mine can be accommodated in a single, mined out longwall panel. (Numbers listed in Table 4.5a are rounded to three significant figures.)

Size of the operation:

```
    1,000 tons of refuse stowed per shift (3000 tons per day)
    3,000 tons per day raw longwall production from a single panel
   10,700 tons per day raw coal produced from the entire mine
    2,700 tons per day refuse produced from the entire mine
2,464,000 tons per year raw coal from the entire mine
1,850,000 tons per year clean coal from the entire mine
  560,000 tons per year refuse from the entire mine backfilled
   56,000 tons per year fine refuse impounded on the surface
```

(1) Crushing and Loading (200 tons per hour crushed to 5" x 0 to 3" x 0)

Capital[a]

Crusher	$ 19,000
Feed conveyor 30 inches x 200 feet	50,000
Loadout bin 150 tons-18 foot Clearance including operators enclosure	95,000
Foundations (21 cubic yards)	3,200
Miscellaneous electrical--95 HP	5,700
Subtotal	$ 172,900
Engineering costs and contingencies (15%)	25,900
TOTAL capital expenditure	$ 198,800

Capital expenditure per year = $198,800 x .2588 CRF = $51,449/year

Labor
1 man/shift at $66 for 230 days = $45,540

Operation, Supplies, and Maintenance

Crusher maintenance	0.002/ton refuse
Belt	0.002
Power	0.004
Miscellaneous--paint, electrical	0.002
	0.01 /ton refuse

$.01 x 560,079 tons refuse/year = $5,601/year

(2) Surface Transportation[b]
3,000 tons of refuse per day is approximately 150 truck loads per day. The minimum charge for contract trucking with highway vehicles for distances greater than one mile and less than 8 miles is $.65/ton. $.65/ton x 560,079 tons/year = $364,051/year

TABLE E(a)   (Continued)

(3)   Storage and Handling
If 420,051 tons refuse per year are produced by the mine excluding the longwall, 1,826 tons of refuse are produced per day.

On those days that the longwall is being moved this refuse must be stored on the surface before being backfilled. This amounts to 1,826 tons of refuse that must be stored during the 43.22 days per year the longwall is being moved. The stockpile for this refuse will be 35,620 tons (20 days of 1,826 tons per day). The longwall panel can accommodate 3,000 tons per day of refuse of which 2,576 tons of refuse will be produced daily from the mine (including the longwall) plus 424 tons per day from the storage area.

(A)   At the Plant

Capital
One 4-cubic yard front-end loader will be required to reload the 35,620 tons of refuse (2.16 tons/year) that must be stockpiled while the longwall is being moved.
$53,500 x .2588 CRF = $13,846/year

Labor
1 man to load 424 tons per day from the stockpile for trucking to the injection site and for clean-up and maintenance around the crusher and bin:
1 man-shift/day $66 x 230 days/year = $15,180

Operation, Supplies, and Maintenance
Cost/day = $30
$30 x 187 days/year = $5,610/year

Contract Trucking
The 76,939 tons per year which is stockpiled (2.16 x 35,620) would be trucked with off-the-road vehicles to a storage area @ $.30/ton of refuse.
$.30 x 76,939 = $23,081/year

(B)   At the Borehole Site

Capital
Truck hopper with stacker belt $105,000
Highlift--10 yard bucket   160,000
    TOTAL capital expenditure $265,000
Capital expenditure per year: $265,000 x .2588 CRF = $68,582/year

Labor
1 man/shift
3 x $66 x 230 days/year = $45,540

TABLE E(a) (Continued)

> Operation, Supplies, and Maintenance
> Diesel fuel, maintenance of high-lift
> $11.00/hour x 21 hours/day x 186 days/year = <u>$42,966/year</u>

(4) 400 foot Borehole and Borehole Site

<u>Capital</u>
A 12-inch borehole, cased to 10 inches would cost $32/foot.
$32 x 400 ft = $12,800 per borehole
It takes 1.5 boreholes per panel (3 boreholes for 2 panels opposite each other) to inject the material to the longwall face
1.5 boreholes x $12,800 = $19,200 per panel
2.16 panels/year x $19,200 = <u>$41,472 for boreholes per year</u>

Borehole sites would have to be purchased or leased
Each site would require:

Negotiations	$ 300
Lease	1,000
Site grading	500
Restoration of property after use	1,000
Access road (1,500' long)($4/ft)	6,000
TOTAL cost per borehole site	$8,800

1.5 boreholes/panel x 2.16 panels/year x $8,800/site = <u>$28,512/year</u>

TOTAL capital expenditure per year for borehole sites and boreholes:
$69,984

(5) Pneumatic Stowing System

<u>Capital</u>

[c]Stower (Brieden KZ 250 peak capacity 450 tons per hour)	$ 32,250
[e]Diesel Motor	8,000
[c]Valve	250
[c]12" elbows: 4 @ $910	3,640
[c]Telescopic sections: 4 @ $1,335	5,340
[c]Hydraulic cylinders for telescopic sections: 8 @ $445	3,560
[c]Hydraulic deflectors: 40 @ $3,950	158,000
[c]Deflector skids: 20 @ $150	3,000
[c]Discharge walls: 80 @ $100	8,000
[c]Spring mounts: 120 @ $60	7,200
[c]Ball joint sections: 3 @ $1,375	4,125
[d]12" roadway pipe: 2,500 ft @ $50/ft	125,000
[d]Extended longwall jacks	312,500
[c]Duties and customs (6%)	40,252
TOTAL capital expenditure	$711,117

Capital expenditure per year: $711,117 x .2588 CRF = <u>$184,037</u>

TABLE E(a)   (Continued)

Labor
11 men-shifts/day at $66 x 230 days/year = $166,980

Operation, Supplies, and Maintenance
[f]Compressor rental and maintenance $120,000
Maintenance of stowing system: 10% of capital   71,387
Fuel   55,000
Tax and insurance   4,000
Pipe maintenance and replacement (150,000 ton life)   350,000
    TOTAL operating cost per year $600,387

(6) Disposal of Fines
     Same as in surface disposal (see Appendix C, Table D, Section 3).

     Capital
     $500,000 x .2588 CRF $129,400/year

     Labor
     50 man-days/year x $66 $ 3,300/year

     Maintenance
     56,000 tons of fines x 1 mile x 0.05/ton mile $ 2,800/year

(7) Productivity Loss
     A 1% loss in productivity to a mining operation would result in a cost of
     $.067/ton of coal or $.20/ton of refuse (based on Consolidation Coal
     mining operations). If 25% of a mine's productivity is longwall pro-
     duction, a 1% loss in production at the longwall will result in a 0.25%
     loss in production from the entire mine. A 0.25% loss in productivity
     to the entire mine will result in a cost of $.017 per ton of coal or
     $0.05/ton of refuse.

Loss of Productivity at the Longwall	Cost/ton refuse	Cost/ton coal
1%	$ .05	$ .017
5%	$ .25	$ .085
10%	$ .50	$ .170
15%	$ .75	$ .255

[a]Pennsylvania Crusher.
[b]Present Contract Trucking Charge in Western Pennsylvania Area.
[c]Singh and Courtney, 1974.
[d]Radmark Engineering, personal communication, 1973.
[e]Caterpillar.
[f]Gardner-Down.

TABLE E(b)   Pneumatic Backfilling (breakdown of Table 4.5b)

Estimate of the cost of pneumatic backfilling behind longwall faces when the amount of waste produced by the mine exceeds the capacity of a single longwall panel. Two longwall panels would be backfilled, one completely, the other with that remaining waste. The backfilling system would not be shut down while one longwall panel moved because the other longwall would accomodate most of its waste.

Size of the operation:
```
        3,333 tons per day waste backfilled (1,111 tons per shift)
        3,000 tons per day longwall production from a single panel
       15,000 tons per day raw coal production from the entire mine
        3,666 tons per day waste produced from the entire mine
    3,373,000 tons per year raw coal from the entire mine
    2,530,000 tons per year clean coal from the entire mine
      766,600 tons per year backfilled
       76,600 tons per year fine waste impounded on the surface
```

(1)   Crushing and Loading
      Same as for Table E(a)

      Capital
      $198,000 x .2588 CRF $ 51,449/year

      Labor
      1 man/shift at $66 for 230 days $ 45,540/year

      Maintenance
      $.01 x 766,600 tons of waste $ 7,666/year

(2)   Surface Transportation
      766,600 tons/year of waste x $.65/ton $498,290/year

(3)   Storage and Handling
      On those days that one longwall panel is being moved, not all the mine's waste will be accommodated behind the operating longwall. 333 tons of waste will be stockpiled while each longwall is moved. (86.4 days)

      (A)   At the Plant

            Capital
            One 4 cubic yard front end loader can reload 333 tons of waste for 86.4 days (28,770 tons).
            $53,500 for caterpiller #966
            $53,500 x .2588 CRF = $13,846/year

TABLE E(b)   (Continued)

Labor
Same as Table E(a)
1 man-shift/day x $66 x 230 days/year = $15,180/year

Operation, Supplies, and Maintenance
Cost/day = $25 x 186.6 days/year = $4,665/year

(B)   At the Borehole Site

Capital
Because the borehole sites will be used, the site costs of Table E(a) will double.

Borehole No. 1
Truck hopper with stacker belt $105,000
10 cubic yard highlift $160,000

Borehole No. 2
Truck hopper with stacker belt $105,000
4 cubic yard highlift $ 53,000

TOTAL capital expenditure for both holes $423,000
$423,000 x .2588 CRF = $109,472/year

Labor
2 men/shift
6 x $66 x 230 days/year = $91,080/year

Operation, Supplies, and Maintenance
Borehole No. 1
$11/hour x 21 hours/day x 186.8 days/year $ 43,150

Borehole No. 2
$100/day x 143.6 days/year $ 14,360
$11/hour x 21 hours/day x 43.2 days/year $ 9,979

TOTAL for both boreholes: $67,489/year

(4)   400 foot Borehole and Borehole Site

Capital
A 400 foot borehole would cost $12,800
6.48 boreholes per year will be required to backfill behind two longwall miners.
Each site costs $8,800 (see Table E(a))
6.48 x $8,800 = $57,024
TOTAL capital expenditure: $139,968

TABLE E(b)   (Continued)

(5)   Pneumatic Stowing System

Capital
Double the cost of Table E(a)
2 x $710,637 = $1,421,274
$1,421,274 x .2588 CRF = $367,826

Labor
Borehole and Longwall No. 1
11 men/day x $66 x 230 days/year = $166,980

Borehole and Longwall No. 2
5 men/day x $66 x 230 days/year = $75,900

TOTAL for both boreholes: $242,880

Operation, Supplies, and Maintenance
Double the cost of Table E(a) $1,200,000

(6)   Disposal of fines
Same as in surface disposal and in Table E(a)

Capital
$500,000 x .2588 CRF $ 129,400/year

Labor
50 man-days/year x $66 $ 3,300/year

Maintenance
76,600 tons of fines x 1 mile x 0.05 ton mile $ 3,830/year

(7)   Productivity Loss
Same as Table E(a)

TABLE E(c) Pneumatic Backfilling (breakdown of Table 4.5c)

Estimate of the cost of pneumatic backfilling behind a longwall face when the amount of waste to be stowed is less than the volume of coal extracted, i.e., only the waste production from the longwall and its associated entries.

Size of the operation:
```
    1,000 tons per day waste backfilled (333 tons per shift)
    3,000 tons per day longwall production from a single panel
    4,400 tons per day raw coal production from the entire mine
    1,100 tons per day waste produced from the entire mine
1,012,000 tons per year raw coal from the entire mine
  759,000 tons per year clean coal from the entire mine
  230,000 tons per year backfilled
   23,000 tons per year fine waste impounded on the surface
```

(1) Crushing and Loading

   Capital
   Same as Table E(a)
   $198,880 x .2588 CRF = $51,449/year

   Labor
   Same as Table E(a)
   1 man/shift x $66 x 230 days/year = $45,540/year

   Operation, Supplies, and Maintenance
   Same as Table E(a)
   $.01 x 230,000 tons of waste = $2,300/year

(2) Surface Transportation
   230,000 ton/year of waste x $.65 = $149,500/year

(3) Storage and Handling
   Because the longwall can easily accommodate all the waste, there are no storage costs at the plant except for handling the waste from developing panels while the longwall is moved.

   (A) At the Plant

      Capital - none

      Labor - none

      Contract Trucking and Loading (10,800 tons/year)
      30¢ to dump at the plant and 60¢ to reload
      $.90 x 10,800 tons/year = $9,720

TABLE E(c)   (Continued)

(B)   At the Borehole

Capital

Truck hopper with stacker belt	$ 65,000
4 cubic yard highlift caterpillar 966	53,500
TOTAL capital expenditure	$118,500

Capital expenditure/year
$118,500 x .2588 CRF = $30,668/year

Labor
1 man/shift at $66 for 230 days/year = $45,540/year

Operation, Supplies, Maintenance
$6.28/hour highlift and stoker operating cost
$6.28/hour x 21 hours/day x 186.8 days/year = $24,635/year

(4)   400 foot Borehole and Borehole Sites

Capital
Same number and costs as Table E(a)
1.5 holes/panel x 2.16 panels/year x $12,800/hole = $41,472/year
Each site costs $8,800 (see Table E(a))
3.24 x $8,800 = $28,512/year
TOTAL capital expenditure per year: $69,984 per year

(5)   Pneumatic Stowing System

Capital

200 ton per hour Brieden system	$ 27,000
The rest same as Table E(a)	678,387
TOTAL capital expenditure	$705,387

TOTAL capital expenditure per year
$705,387 x .2588 CRF = $182,554

Labor
7 men/day x $66 x 230 days/year = $106,260

Operation, Supplies, and Maintenance
Same as Table E(a)   ($599,000) except half the pipe replacement
$450,000 per year

(6)   Disposal of Fines
Same as in surface disposal and in Table E(a)

Capital
$500,000 x .2588 CRF = $129,400/year

Labor
50 man/days/year x $66 = $ 3,300/year

TABLE E(c)   (Continued)

Maintenance
23,000 tons of fines x 1 mile x 0.05/ton mile = $1,150/year

(7)   Productivity Loss
Because there is no necessity for the backfilling system to maintain pace
with the mining and transport system, the loss in productivity of the
miner will not be as affected by the performance of the stower as in
Tables E(a) and E(b). A conservative estimate of productivity loss is
one hour per day or 5%.

TABLE F Committee Estimate of the Cost of Hydraulic Backfilling (breakdown of Table 4.6 from Chapter 4)

The cost estimates of Table 4.6 are based on the following costs rounded to three significant figures:

Size of the operation:
     3,000 tons of refuse backfilled per day (1,000 tons per shift)
   690,000 tons of refuse backfilled per year
2,070,000 tons of clean coal produced per year

(1) Crushing & Loading (-3/4" Crushing - 200 TPH)

Capital

Belt--Plant to crusher	$ 50,000
Belt Magnet	6,000
Feed Belt	34,000
Screens	18,200
Hammermill C-100-42	29,500
Motor--300 HP	10,900
Conveyor for Refuse and Scrap	6,300
150-ton Bin	95,000
Building	35,500
Miscellaneous Electrical	30,800
Miscellaneous Chute Work	20,000
Foundation	10,500
	$ 346,750
15% Engineering costs and Contingencies	52,013
TOTAL capital expenditure	$ 398,763

Capital expenditure per year $398,763 x .2588 CRF = $103,200/year

Labor
3 men/day x $66 x 230 days/year = $45,540

Operation, Supplies and Power	$/Ton Refuse
Crusher Maintenance	$ .075
Screener	.01
Conveyors	.002
Power	.02
Miscellaneous	.005
	$ .112

$.112 x 690,000 tons/year = $77,280/year

(2) Surface Transportation

Contract trucking a distance of 4 miles
$.65/ton x 690,000 tons = $448,500/year

TABLE F (Continued)

(3) Borehole and Site
     690,000 tons of refuse stowed/year
     50,000 tons put down each borehole = 15 boreholes/year
     Boreholes drilled to 10" and cased to 8" at $20/foot
     Section to be filled at 400-foot depth

     Capital
     400 ft borehole at $20/foot $ 8,000/borehole
     Additional costs (land acquisition, roads,
       etc. see Appendix C, Table E(a), Section (4))   8,300/borehole
     Working Capital   1,500/borehole
     TOTAL capital cost $17,800/borehole
     15 boreholes/year at $17,800/borehole = $267,000/year

     Labor
     1 man/day
     1 x $66/day x 230 days/year = $15,180

     Operation, Supplies, and Maintenance
     Moving bin, etc. from site to site at $200/borehole site
     $200/borehole x 15 boreholes/year = $3,000/year

(4) Surface Water Supply System
     15,000 feet of 8-inch water line from the preparation plant to the
     borehole, and to recirculate the water from the backfilled section.

     A hopper, blender, and pumping unit on the surface (such as the mobile
     unit used in Rock Springs, Wyoming).

     Capital
     Pump (2,000 gpm) $ 80,000
     Hopper and conveyor loader   70,000
     Totally mobile stowing unit: mixing tub, blender,
       pump (5,000 gpm), gear box, and control area   125,000
     TOTAL capital expenditure $275,000
     Cost per year: $275,000 x .2588 CRF = $71,170/year
     15,000 feet of 8-inch water line at $17/foot = $255,000/year
     TOTAL capital expenditure per year = $326,170/year

     Labor
     3 men/day at $66 x 230 days/year = $45,540

     Operation, Maintenance and Power
     Power: 300 HP at 15 hours/day x 230 days/year at $.015/kWh = $11,500/
              year
     35,000 gallons of diesel fuel = $7,000
     Parts and Maintenance: 10% stowing unit's capital cost = $12,500/year
     TOTAL expenditure per year = $24,000/year

TABLE F   (Continued)

(5)   Underground Pumping & Recirculation System

<u>Capital</u>

Stainless steel pump for 1,500 gpm, 400 foot head	$ 60,000
Motor, switch gear, transformer, cable	60,000
1,000 feet of pipe at $12/foot	12,000
Miscellaneous small pumps and pipe and electrical equipment for 600 gpm	60,000
TOTAL capital expenditure	$192,000

$192,000 x .2588 CRF = $49,690/year
2 boreholes for the recirculating water at $17,800 = $35,600/year
TOTAL capital expenditure/year = <u>$85,290</u>

<u>Labor</u>
2 men/day at $66 for 230 days/year = <u>$30,360</u>

<u>Operation, Maintenance and Power</u>
Maintenance and Parts 10% of capital = $8,529/year
Power 300 HP at 15 h/day at $.015/kWh = $11,500/year
TOTAL expenditure per year = $20,029

(6)   Underground Storage Site

<u>Capital</u>
Cement block and mortar $50/bulkhead
7 bulkheads/panel, 22.4 panels/year = 156.8 bulkheads/year
156.8 bulkheads/year x $50 = <u>$7,840/year</u>

<u>Labor</u>
$380/bulkhead
$380/bulkhead x 156.8 bulkheads/year = <u>$59,584/year</u>

(7)   A 1% loss in productivity to the entire mine would result in a cost of $.20/ton of refuse or $.067/ton of coal.

Water has a devastating effect on mine productivity (see text of chapter three). Unless the disposal area were located in a totally abandoned area of a mine distant from the working faces and haulageways, the productivity loss to the mine would be at least 10%.

Overall Mine Productivity Loss	Cost/ton refuse	Cost/ton Coal
5%	$1.00	$ .33
10%	$2.00	$ .67
15%	$3.00	$1.00

APPENDIX D: EXCERPTS FROM REGULATIONS REGARDING THE
DISPOSAL OF COAL MINE WASTE

U.S. REGULATIONS

As of January 1, 1974 the Federal Regulation regarding the disposal of coal mine waste read as follows:

77.214   Refuse piles; general.

a.   Refuse piles constructed on or after July 1, 1971, shall be located in areas which are a safe distance from all underground mine airshafts, preparation plants, tipples, or other surface installations and such piles shall not be located over abandoned openings or steamlines.
b.   Where new refuse piles are constructed over exposed coal beds the exposed coal shall be covered with clay or other inert material as the piles are constructed.
c.   A fireproof barrier of clay or inert material shall be constructed between old and new refuse piles.
d.   Roadways to refuse piles shall be fenced or otherwise guarded to restrict the entrance of unauthorized persons.
36 F.R. 9364, May 22, 1971, as amended at
36 F.R. 13143, July 15, 1971.

77.215   Refuse piles; construction requirements.

a.   Refuse deposited on a pile shall be spread in layers and compacted in such a manner so as to minimize the flow of air through the pile.
b.   Refuse shall not be deposited on a burning pile except for the purpose of controlling or extinguishing a fire.
c.   Clay or other sealants shall be used to seal the surface of any refuse pile in which a spontaneous ignition has occurred.
d.   Surface seals shall be kept intact and protected from erosion by drainage facilities.
e.   Refuse piles shall not be constructed so as to impede drainage or impound water.
f.   Refuse piles shall be constructed in such a manner as to prevent accidental sliding and shifting of materials.
g.   No extraneous combustible material shall be deposited on refuse piles.

77.216   Retaining dams: construction; inspection; records.

a.   If failure of a water or silt retaining dam will create a hazard, it shall be of substantial construction and shall be inspected at least once each week.
b.   Weekly inspections conducted pursuant to paragraph (a) of this 77.216 shall be reported and the report shall be countersigned by any of the persons listed in paragraph (d) of 77.1713.

U.S. PROPOSED REGULATIONS

DEPARTMENT OF THE INTERIOR

Mining Enforcement and Safety Administration
[30 CFR Part 77]
Mandatory Safety Standards, Surface Coal Mines
and Surface Work Areas of Underground Coal Mines
Notice of Proposed Rule Making

Notice is hereby given that in accordance with the provisions of section 101(a) of the Federal Coal Mine Health and Safety Act of 1969 (P.L. 91-173), and pursuant to the authority vested in the Secretary of the Interior under section 101(a) of the Act, it is proposed that Part 77, Subchapter 0 of Chapter I, Title 30, Code of Federal Regulations be amended by adding paragraphs (h) and (i) to §77.215 and by adding §§77.215-1 through 77.215-4, and by deleting present §77.216 and inserting in lieu thereof §§77.216 through 77.216-3 as set forth below.

The proposed amendments to section 77.215 and new §§77.215-1 through 77.215-4 set forth requirements for the construction of refuse piles and extinguishment of fires; reporting pertinent information on refuse piles; certifying its stability; abandonment; and identification of the pile.

The new §§77.216 through 77.216-3 will require development and approval of plans in constructing structures which impound water or silt; inspection of impoundments; notification of potentially hazardous conditions; and identification of an impoundment.

Interested persons may submit written comments, suggestions or objections to the Administrator, Mining Enforcement and Safety Administration, Room 4513, Interior Building, 18th and C Streets, N.W., Washington, D.C. 20240, no later than March 1, 1974.

(January 11, 1974)   (Sgd) C.K. Mallory
                       Deputy Assistant Secretary of the Interior

Part 77, Subchapter 0, Chapter I, Title 30 Code of Federal Regulations will be amended as follows:

1. §77.215 will be amended by adding new paragraphs (h) and (i) thereto and new §§77.215-1 through 77.215-4 will be added as follows:

§77.215 Refuse piles; construction requirements.

(h) On and after the effective date of this paragraph, new refuse piles or refuse deposited on existing piles shall be compacted in layers of a maximum of 2 feet in height and shall have a slope no greater than 27°.
(i) Fires in refuse piles shall be extinguished by sealing or excavation of the burning material.

§77.215-1 Refuse piles; reporting requirements.

(a) Within 180 days following the effective date of this section, the operator of a coal mine on whose property a refuse pile is located shall make an accurate survey of such pile and the area extending 500 feet around its perimeter, if such pile can present a danger to miners on mine property, and submit to the Coal Mine Health and Safety District Manager for the district in which the refuse pile is located a report showing as a minimum:

(1) the location of the refuse pile shown on the USGS topographic quadrangle map of the largest scale available;
(2) such construction history as is available;
(3) at an adequate scale for the site and pile, cross sections of its length and width at the point of maximum depth showing the approximate original ground surface and present configuration with mean sea level elevations at significant points;
(4) whether or not the refuse pile is burning and, if so, the measures being taken to extinguish the fire;
(5) whether or not the refuse pile is in active use;
(6) proximity to nearest stream and elevation of the toe above such stream if the stream could foreseeably have an adverse effect on the stability of the pile;
(7) location of diversion drains and other facilities to ensure that water is not impounded;
(8) any other information required by the Coal Mine Health and Safety District Manager for the district in which the refuse pile is located.

(b) After the initial report required by paragraph (a) of this section a report shall be submitted annually for refuse piles in active use and every three years for those which have been abandoned.

§77.215-2 Refuse piles; certification of stability.

Within 180 days following the effective date of this section, any refuse pile required to be reported under 77.215-1(a), shall be certified as being safe by a registered engineer with a knowledge of mine and processing plant refuse, soil mechanics, and hydrology; or, in lieu thereof, a report shall be submitted indicating what additional investigations, analyses, or improvement work is necessary before such a certification can be made, including what provisions have been made to carry out such work in addition to a definitive schedule for completion of such work.

The certification or report shall be submitted to the Coal Mine Health and Safety District Manager for the district in which the refuse pile is located and shall contain the statement that "the refuse pile cannot impound water and is constructed in such a manner as to preclude the probability of failure of such magnitude as to endanger the lives of coal miners." After the initial certification required by this section, certifications shall be submitted annually for all refuse piles in active use and every three years for those which have been abandoned. Certifications shall include copies of all information considered in making the analysis.

§77.215-3 Refuse piles; abandonment.

Prior to permanent abandonment of any refuse pile reported under the requirements 77.215-1(a), the operator shall submit to and obtain approval of the Coal Mine Health and Safety District Manager for the district in which the refuse pile is located, a plan for abandonment which shall contain provisions to preclude future impoundment of water and major slope instability.

§77.215-4 Refuse piles; identification.

Within 180 days following the effective date of this section, the operator of a coal mine on whose property a refuse pile reported under the requirements of 77.215-1(a) is located shall place an identification marker, at least six feet high and of permanent construction, on, or immediately adjacent to, each refuse pile and he shall include the site name, the site owner, and the site number on each marker in accordance with the identification system established by the Coal Mine Health and Safety District Manager for the district in which the refuse pile is located.

2. Present §77.216 will be deleted and a new §77.216 will be inserted in lieu thereof and new §§77.216-1 through 77.216-3 will be added as follows:

§77.216 Water or silt impoundments; requirements for approval of plans.

(a) Plans for impounding water or silt shall be required under this section if an existing or proposed structure, which presents a hazard to coal miners on coal mine property, can:

  (1) impound water or silt to a height of five feet or more above the downstream toe of the embankment and can have a storage volume of 20 acre-feet or more; or,
  (2) impound water or silt to a height of 20 feet or more above the downstream toe of the embankment; or,
  (3) present a hazard to coal miners regardless of storage volume.

(b) All new water or silt impoundments which meet the requirements of paragraph (a) of this section and are located on coal mine property shall be designed and constructed by a registered engineer in accordance with an approved plan. Within 180 days following the effective date of this section, the operator of any mine with any water or silt impoundments which meet the requirements of paragraph (a) of this section and which were in existence prior to the effective date of this section shall submit for approval a plan prepared by a registered engineer for the continued use or final abandonment of such structure.

(c) The registered engineer shall have a knowledge of mine and processing plant refuse, soil mechanics, and hydrology, and experience in earth dams design and construction. Plans shall be submitted to and approved by the Coal Mine Health and Safety District Manager for the district in which the structure is located and shall include a certification by the registered engineer responsible for design and construction of the impoundment that: "the structure is safe for the maximum volume of water or silt which can be impounded therein;" except that in the case of new structures the certification shall be made after the structure can impound water or silt to the degree specified in paragraph (a) of this section; or in lieu thereof, a report shall be submitted indicating what additional investigations, analyses, or improvement work is necessary before such a certification can be made, including what provisions have been made to carry out such work in addition to a definitive schedule for completion of such work. After the initial certification required by this section, certification shall be submitted annually thereafter as long as the structure can impound water or silt to the degree specified in paragraph (a) of this section.

§77.216-1 Water or silt impoundments; minimum plan requirements; changes or modifications; abandonment.

(a) The plan as specified in §77.216, shall contain as a minimum the following information:

(1) the purpose for which the structure is or will be used;

(2) an accurate survey of the site showing the existing or proposed location of the structure and the watershed contributing to the impoundment illustrated on a USGS topographic quadrangle map of the largest scale available;

(3) the type of foundation on which the structure is or will be constructed;

(4) the type, size range, and physical and engineering properties of the materials used, or to be used, in constructing each zone or stage of the structure; the method of site preparation and construction of each zone; the approximate dates of construction as is available, and any record or knowledge of embankment failures;

(5) detailed drawings of the structure, including plan, cross-section, and profile views, showing all zones, foundation improvements, drainage provisions, spillways, diversion ditches, outlets, instrument locations, slope protection, and other elements needed for embankment design, in addition to the present, or proposed, initial and final freeboard, siltation level, water level and other information pertinent to the impoundment itself, including any identifiable natural or manmade features which could affect operation of the impoundment;

(6) the type and purpose of existing or proposed instrumentation, if any;

(7) the area, maximum depth, and maximum volume of the embankment and silt or water that can be impounded;

(8) the runoff attributable to a maximum probable flood;

(9) the runoff attributable to the storm for which the structure is designed;

(10) the spillway and diversion design features and capacities;

(11) computed factor of safety range, methods used to determine the range, and all data or assumptions on which these computations were based;

(12) the proximity of the structure to the nearest stream;

(13) the locations of surface and underground coal mine workings including the depth and extent of such workings in the interval from the upstream extent of the impoundment to the first major stream downstream of the structure;

(14) estimated depth of flood, in feet, as well as quantity and rate of flow at the first location where miners may be endangered in the event of a possible breach in the structure at the maximum possible storage level;

(15) a copy of the most recent topographic map available showing the flood plain downstream of the impoundment to the point where the drainage area meets a river having an average capacity of about 50 percent of the expected failure flood flow;

(16) provisions for maintenance and repair of the structure;

(17) provisions for abandonment;

(18) such other information as may be required by the Coal Mine Health and Safety District Manager for the district in which the structure is located.

(b) Every six months following the submission of information specified in paragraph (a) of this section, the operator shall submit to the Coal Mine Health and Safety District Manager for the district in which the structure is located, a report defining any changes in the embankment geometry, instrumentation, average and maximum depths and elevations of the impounded silt and water, storage capacity of the structure, the volume of silt or water impounded, and any other aspect of the structure affecting its factor of safety which has occurred during such reporting period.

(c) Any changes or modifications to water or silt retaining structures other than those stipulated in the plan approved under Section 77.216 shall be approved by the Coal Mine Health and Safety District Manager for the district in which the structure is located prior to the initiation of such changes or modifications.

(d) Prior to permanent abandonment of any water or silt impoundment, the operator shall submit to and obtain approval of the Coal Mine Health and Safety District Manager, for the district in which the structure is located a plan for abandonment which shall contain provisions to preclude future impoundment of water or silt, or major slope instability.

(e) The information required in paragraphs (a), (b), (c), and (d) of this section shall be kept and certified by the principal official in charge of health and safety at the mine as designated by the operator in accordance with Section 107 (d) of the Federal Coal Mine Health and Safety Act of 1969.

§77.216-2 Water or silt impoundments; inspection requirements; correction of hazards; program requirements.

(a) All water and silt impoundments which meet the requirements of 77.216(a) shall be examined at least once weekly for appearances of structural weakness, volume overload and other potentially hazardous conditions and all instruments shall be monitored at the approved time intervals by a qualified person designated by the operator. When rising water or silt approaches the safe design capacity of the structure, examinations shall be made by a qualified person designated by the operator at least once every eight hours, or more often as required by an authorized representative of the Secretary, and the Coal Mine Health and Safety District Manager for the district in which the structure is located shall be notified.

(b)   When a potentially hazardous condition exists the operator shall immediately:

(1)   notify the Coal Mine Health and Safety District or Subdistrict Manager for the district or subdistrict in which the structure is located;
(2)   eliminate the potentially hazardous condition;
(3)   notify and prepare to evacuate, if necessary, all coal miners within the area which may be affected by the potentially hazardous conditions.

(c)   Records of the inspections required by paragraph (a) of this section, including instrumentation monitoring, shall be kept and certified by the principal official in charge of health and safety at the mine, as designated by the operator in accordance with Section 107 (d) of the Federal Coal Mine Health and Safety Act of 1969. Copies of such records shall be available at the mine for inspection by an authorized representative of the Secretary.

(d)   The operator of each coal mine with a water or silt impoundment meeting the requirements of 77.216(a) shall adopt a program for carrying out the requirements of paragraphs (a) and (b) of this section. The program shall be submitted for approval to the Coal Mine Health and Safety District Manager for the district in which the structure is located, within 180 days following the effective date of this section. The program shall include as a minimum:

(1)   a schedule and procedures for inspection of the impoundment by a designated qualified person;
(2)   a schedule and procedures for monitoring any required or approved instrumentation by a designated qualified person;
(3)   procedures for evaluating potentially hazardous conditions;
(4)   procedures for notifying the appropriate Coal Mine Health and Safety District or Subdistrict Manager;
(5)   procedures for eliminating the potentially hazardous conditions;
(6)   procedures for removing all miners from the area which may be affected by the potentially hazardous conditions; and
(7)   any additional information which may be required by the Coal Mine Health and Safety District Manager for the district in which the structure is located.

(e)   Before making any changes or modifications in the program approved in accordance with paragraph (d) of this section, the operator shall obtain approval of such changes or modifications from the Coal Mine Health and Safety District Manager for the district in which the structure is located.

(f) The qualified person referred to in paragraphs (a), (d)(1), and (d)(2) of this section shall be trained to recognize specific signs of embankment instability, and other potentially hazardous conditions, by visual observation and, if applicable, instrumentation monitoring.

§77.216-3 Water or silt impoundments; identification.

Within 180 days following the effective date of this section, the operator of a coal mine on whose property a water or silt impoundment meeting the requirements of 77.216(a) is located shall place an identification marker, at least six feet high and of permanent construction on or immediately adjacent to each impoundment and he shall include the site name, the site owner, and the site number, on each marker in accordance with the identification system established by the Coal Mine Health and Safety District Manager for the district in which the structure is located.

BRITISH MEMORANDUM

The following excerpt from the Report of the Tribunal Appointed to Inquire into the Disaster at Aberfan on October 21, 1966 summarizes British objections to backfilling for waste disposal. (Mechanized lateral discharge systems for pneumatic backfilling have not been used in Britain.)

NATIONAL COAL BOARD MEMORANDUM
ON UNDERGROUND STOWING

1. Before and since nationalisation the *desirability* of stowing mine waste underground in the Colliery being operated has been clearly recognised by the coal mining industry. There has been and still is much consideration given to the multiplicity of problems involved and many experiments have been carried out in relation to run of mine rubbish and washery discard (including tailings).

2. In *general* terms underground stowing is technically feasible, though the problems vary from colliery to colliery and even from seam to seam *but the cost is too great to be justified*. The increased rate at which the coal face must advance if advantage is to be taken of modern mining methods makes it more difficult than ever to apply, economically, a system of stowing large quantities of the waste underground. There has to be considered *not only the actual cost of stowing* but the decrease in rate of output of coal which must necessarily follow. *If underground stowing were* generally *employed the result would be a financial burden that the coal industry could not bear.*

3. Stowing, albeit technically feasible, *creates hazards which cannot be ignored. Airborne dust* is increased and in South Wales in particular this can bring an added risk of *pneumoconiosis*.

4. The German coal industry has experienced problems that are similar to those experienced in the United Kingdom. In the *Ruhr* there has been a marked decline in underground stowing in recent years. Whereas in 1958 about 50 percent of all coal was obtained in conjunction with solid stowing today the figure has been reduced to 16 per cent.

5. *Ten years ago* the Parliamentary Secretary to the Ministry of Housing and Local Government stated in the House of Commons that "there has been the idea of underground stowage of the spoil material, but very often that is not practically possible." (See Hansard for 23rd May, 1957). With greater mechanisation since 1957 and more intensive mining methods *underground stowing is today even less of a practical proposition.*

6. At the present time stowing is employed in some exceptional cases mainly to improve the conditions of faces and underground roadways and to minimise surface subsidence where normal methods of strata control are inadequate. In South Wales stowing is being practised in approximately 2 1/2 per cent of the working faces.

7. *The conclusion is that underground stowing is not economically practicable though technically feasible.* Modern mining methods entail multi-shift working of rapidly advancing, highly mechanised faces. Under these conditions stowing is *a practical impediment* to the type of mining necessary to make the industry economically viable.

8. The underground stowing of tailings has been the subject of special study and experiment, but (a) it has not yet proved possible successfully to avoid the clogging of the stowing pipes, and (b) even if this problem could be overcome in the future it would *not* be economically practicable to stow tailings for the reasons set out above.   (7th March, 1967)

GERMAN REGULATIONS

Extract from the German (North Rhine-Westfalis) Guidelines
regarding the "official authorization of mine waste dumps within
the domain of the Mine Inspection." (1972)

This publication contains a list of rules and regulations related to waste dumps. Some of the more specific items are translated below. NOTE: This is not an authorized, official translation, and is included for general information only.

1.2	Mine waste dumps in the intent of these regulations are embankments, subjected to the Mine Inspection, constructed on the earth surface or protruding above the original ground level when within the limits of surface mines or rock or soil quarries, and build entirely or predominantly out of mine rock or waste from mineral preparation.
2.2	In order to allow a methodical evaluation of the operations plan for the construction or enlargement of a mine waste dump it should, in general, provide the following information:
2.23	Plans of the dump on an appropriate scale; on these should be indicated the growth of the dump in space and in time as well as the diversion of surface water--including all (individual) protective measures at every step. The planned final form and the use of the dump following its completion must be given on a separate plan (landscape plan). The mine operator must hire a landscape architect for the preparation of the landscape plan.

2.24   Composition of the waste to be dumped, especially the amount of flammable products and the amount of products sensitive to water and damaging to plants.

2.25   Specification of the stability of the dump, eventually indication, especially with respect to the foundation (subsurface), of the inclination and composition of the bottom layer as well as composition of the waste; indication of safety distances.

2.26   Description and drawings of the machines and dumping equipment as well as compaction of the dump and measures for the safety of people working on the dump.

2.27   Specification of what the influence on surface and groundwater will be.

2.28   Indications of the type and quantity of the emissions to be expected as well as installations to be used for limiting emissions; as well as known the load at ground level in the area affected by emissions should be given.

2.29   Indications of the measures taken to prevent fires and for their early detection.

2.2.0   Indications of the use that will be made of the topsoil.

2.3, 4, 5, 6, 7   List other agencies and laws directly or indirectly related or involved with problems caused by dumps.

3.   Evaluation of the Operations Plan.

3.1   Basic approach.

3.2   Location of the dump.

3.3   Form (seaping) of the dump.

3.4   Deposit of the fill.

3.5   Water.

3.6   Protection of the neighborhood.

3.7   Reclamation.

3.11   Prior to authorization it must be evaluated whether underground backfill or dumping in mined out sections of open pits or quarries is possible, taking into account the public interest as well as the interest of the mine owner.

3.221   The landscape should be affected as little as possible by the dump.

3.223   Wherever possible a dump should be built upon land with little value for agriculture or forests.

3.31   The final form of the dump must be appropriate for its planned use (landscape plan). In densely populated areas the creation of recreation areas should be attempted (preferred).

3.34   Dumps of material susceptible to spontaneous ignition must, in order to reduce the danger of fires, be shaped so that the wind from the predominant wind direction (must be determined prior to dumping) has the least possible effect. For every single case the danger of spontaneous ignition must be tested.

3.35   The slopes of dumps susceptible to self-ignition must be subdivided into benches. The height of the lowest bench should not exceed 12 meters, the height of the others should in general not exceed 8 meters. For dumps not susceptible to spontaneous ignition it should be decided for every individual case whether and eventually how the slopes should be subdivided.

3.36   The width of the benches should be sufficient to permit a traffic lane (about 4 m) and water runoff towards the dump must be possible.

3.41   Prior to dumping the topsoil and, if present, any more ground layers that can be used for agriculture, must be removed in sufficient quantities to assure the reclamation of the final dump slopes, unless other circumstances, for example protection of groundwater, prevent it.

3.513   The level of the groundwater table as well as the flow direction and water composition must be determined. The thickness of aquifers and the location of public and private water production installations in the area must be determined.

3.637   Supervision of the dump by walking, and if indications of fire risk exist by measurements of temperature and CO, so that timely remedial action is possible.

APPENDIX E: BIOGRAPHIC SKETCHES OF COMMITTEE MEMBERS

Charles Fairhurst, chairman of the committee, Ph.D. in mining engineering, Sheffield University, England, Age: 43, professor and head of the Department of Civil and Mineral Engineering of the University of Minnesota.
Field of expertise:
    rock mechanics
    mine ventilation
    use of underground space
Ongoing Projects:
    ground disturbance due to underground mining
    determination of stresses in rock
    rock bolting of strata in underground mines
    stability of underground openings
Other related experience:
    practical mining experience (in British mines: including pneumatic back-filling and putting out underground mine fires)
    Chairman of the U.S. National Committee for Rock Mechanics
    drilling and blasting
    extensive foreign experience

Joseph P. Brennan, MBA from American University, Age: 38, Vice president for economics and planning, National Coal Association.
Field of expertise:
    the economics of the coal industry
    the social environment of the coal industry
Ongoing projects:
    lecturer in marketing, American University
    the economic and environmental costs of surface mining and underground mining of coal
    surface mining regulation
    coal availability
Other related experience:
    coal research and marketing with the United Mine Workers of America for 16 years, assistant director of research and marketing
    the investment procedures of the coal industry
    knowledge of foreign coal mining industry and economics

Steven L. Crouch, Ph.D. in Mineral Engineering, University of Minnesota, Age: 31, Assistant professor of the Department of Civil and Mineral Engineering, University of Minnesota.
Field of expertise:
    rock mechanics
    stress and strain relationships in flat-lying seams
Ongoing projects:
    computer simulation of caving, ground control, and coal bumps
    the nature and causes of coal bumps and roof falls
    computer applications in mineral engineering

Steven L. Crouch, (continued)
Other related experience:
    mining research in the Republic of South Africa
    stress analysis of the effect of single seam extraction on the associated
       strata

Thomas V. Falkie, Ph.D. in Mineral Engineering, Pennsylvania State University,
Age: 39, Director of the Engineering Department, the Pennsylvania State
University.
Field of expertise:
    mining cost analysis and engineering economics
    mine systems engineering
Ongoing projects (Penn State):
    management theory and practice applied to mining
    land reclamation and disposal of mine wastes
    industrial engineering as applied to mineral operations
    economic analysis
Other related experience:
    simulation of underground mine haulage
    systems evaluation and simulation of mining systems
    survey of capital investment procedures actually used by the mining
       industry
    consultant to the United Nations on mining economics and mine management

Lawrence A. Garfield, degree of Engineer of Mines, Colorado School of Mines,
Age: 53, Project Manager for Viro Dyne Corporation and the Kellogg Corpora-
tion, Denver.
Field of expertise:
    mining, mine maintenance, mine management
    mine research and equipment research
Ongoing projects:
    engineering and management activities in the underground construction
       and mining industries
    designing mines for the extraction of oil shale which include under-
       ground disposal of the spent shale
    teaching a course in mine management
Other related experience:
    extensive mining experience (1941-1969)
    while at White Pine Copper Company (sedimentary copper deposit in
       Michigan) backfilled their room and pillar operations with mine waste
    wide experience with sedimentary deposits other than coal, e.g., gypsum,
       placer gold, sedimentary copper, oil shale

James E. Gilley, degree in Mining Engineering, Virginia Polytechnic Institute,
Age: 45, Division of the Environment, Bureau of Mines, Washington, D.C.
Field of expertise:
    the effects of mining on the environment
    coal mining and coal resources
    Federal regulation, inspection, and enforcement of coal mining activities

James E. Gilley, (continued)
Ongoing projects:
    inventorying and delineating mine-related pollution sources
    surface mine regulations
    remote sensing of waste piles and subsidence detection
Other related experience:
    mine experience (mine construction foreman, production foreman)
    associated with the flushing projects of the abandoned mines for subsi-
      dence control in the Scranton-Wilkes Barre, Pennsylvania area and
      Rock Springs, Wyoming.

Richard E. Gray, degree in Civil Engineering, Carnegie-Mellon University, and
graduate study in geology, Age: 39, Vice president, General Analytics,
Incorporated, Monroeville, Pennsylvania.
Field of expertise:
    engineering geology
    soil and rock mechanics
    groundwater
Ongoing projects:
    contracted by the Corps of Engineers to oversee the detailed engineering
      inspection in Pennsylvania and Maryland of the stability of coal waste
      piles and impoundments
    project manager, state of the art of subsidence control contracted by
      the Appalachian Regional Commission
Other related experience:
    numerous technical papers on stability, subsidence and subsurface
      explorations
    grouting and flushing for subsidence control using materials other
      than coal mine waste
    knowledge of foreign mining and subsidence control practices

J. Davitt McAteer, J. D. West Virginia University College of Law, Age: 30,
Lawyer in the Safety Division, United Mine Workers of America.
Field of expertise:
    law (responsive law, mining law)
    mine health and safety
    the social environment of Appalachia
Ongoing projects:
    represent the UMW in court and at hearings
    teaches mining law at West Virginia University, College of Law
Other related experience:
    Author: *Coal Mine Health and Safety in West Virginia,* (New York:
      Praeger Press) 689 pp.
    Other articles on pneumoconiosis, the safety of some waste impoundments,
      and the social impacts of the coal industry

Walter N. Heine, M.S. in Sanitary Engineering, University of Michigan, Age:
40, Associate Deputy Secretary for Mining and Land Use Planning of the
Department of Environmental Resources of Pennsylvania.

Walter N. Heine, (continued)
Field of expertise:
    state regulation of mining--regulation and enforcement pertaining to
      water pollution, subsidence, surface mining, mine safety and solid
      waste disposal
Ongoing projects:
    chairman, Pennsylvania's commission to study alternative methods of
      disposal of coal mine waste
    research on mine subsidence
    proposed Federal surface mine control legislation
Other related experience:
    mine drainage, its properties and effects
    surface mining rehabilitation technology
    commissioner to the Interstate Mining Compact from Pennsylvania

David R. Maneval, M.S. in Chemistry and Ph.D. in Mineral Preparation, the
Pennsylvania State University, Age: 45, Science Advisor for the Appalachian
Regional Commission.
Field of expertise:
    coal chemistry, utilization, and preparation
Ongoing projects:
    comprehensive study of subsidence control
    comprehensive environmental pollution abatement plan for the Monongahela
      River Basin
Other related experience:
    Pennsylvania Department of Environmental Resources, air and water
      pollution abatement, rehabilitation of surface mined lands
    teaching of mineral preparation
    extinguishment of burning coal refuse banks
    the utilization of mine refuse (antiskid material, construction fill,
      etc.)
    the disposal and utilization of coal mine waste in foreign countries

William N. Poundstone, B.S. in engineering, graduate of the management program
for executives, University of Pittsburgh, Age: 48, Executive Vice president,
Consolidation Coal Company.
Field of expertise:
    coal mining and mine management
    mining equipment and design
    mine transport
Ongoing projects:
    mining research
    alternative transport systems for coal extraction
    gas hazards in coal mines
    design and planning of new mines
Other related experience:
    extensive coal mining experience (1949-   )
    all aspects of coal mining
    foreign experience with coal mining and the disposal of mine wastes

Lee W. Saperstein, D. Phil. in Engineering Science Oxford University, Age: 30, Associate at the Pennsylvania State University.
Field of expertise:
    mining engineering and mining techniques
    materials handling and reclamation
    drilling and blasting
Ongoing projects:
    aspects of noise generation and hearing protection in underground coal
        mines
    planning methods and costs for integrated reclamation and surface mining
        of coal
Other related experience:
    practical mining experience in Europe and the United States
    ventilation research
    stress and strain analysis: Angle of internal friction of mechanically
        chipped material
    rehabilitation of surface mined land and waste piles